QUANTUM LEAP

HOW QUANTUM COMPUTING WILL SHAPE THE FUTURE OF TECHNOLOGY, BUSINESS, AND EVERYDAY LIFE

DR. OLADIPUPO JUDE MESOLE

DISCLAIMER

This book is intended for informational and educational purposes only. While every effort has been made to ensure the accuracy and completeness of the information provided, the author and publisher assume no responsibility for errors, omissions, or any outcomes resulting from applying the concepts and techniques discussed.

The field of quantum computing is rapidly evolving, and discoveries, technologies, and interpretations may render some content outdated or subject to revision. Readers should consult additional sources and professionals for the latest developments and personalized advice.

The examples, case studies, and scenarios presented in this book are hypothetical or based on general knowledge and should not be construed as specific recommendations or endorsements. Any decisions or actions taken based on the information in this book are the reader's sole responsibility.

The author and publisher disclaim any liability for any damages or losses, including but not limited to financial, technical, or personal consequences, arising from the use or reliance on the content herein.

By continuing to engage with this book, readers acknowledge

and accept these terms and agree to use the information at their discretion.

TABLE OF CONTENT

INTRODUCTION

Imagine a future where diseases like cancer are no longer death sentences. Instead, quantum computing accelerates the development of personalized treatments, enabling the analysis of complex genetic data in seconds. Picture a world where supply chain crises are avoided as quantum algorithms dynamically optimize global logistics. These transformative scenarios are not science fiction but glimpses of a revolution led by quantum computing.

Quantum computing is no longer the exclusive domain of physicists and computer scientists. It is rapidly evolving from a theoretical possibility to a practical application, promising to reshape technology, business, and everyday life. This book aims to demystify quantum computing, making it accessible to a broad audience while exploring its transformative potential across diverse fields. By the end of this journey, you will clearly understand how quantum computing works, its challenges, and the profound ways it can impact our world.

WHO SHOULD READ THIS BOOK

This book is for anyone intrigued by the quantum revolution. Whether you are a tech enthusiast eager to stay ahead of the curve, a business leader looking to leverage cutting-edge advancements, a student exploring career opportunities, or a curious mind seeking to understand the future, this book has something for you. No prior physics or computer science expertise is required—just a willingness to dive into qubits, superposition, and entanglement.

SETTING THE STAGE

To appreciate the significance of quantum computing, it's helpful to consider the limits of classical computers. Classical computers, which power our phones, laptops, and data centers, have served us well. They process information in bits, which are binary units representing either a zero or a one. However, classical computing is nearing its limits as our needs grow, from modeling climate change to developing more efficient drugs. Enter quantum computing, a technology that leverages quantum mechanics principles to solve previously deemed unsolvable problems.

WHAT THIS BOOK WILL COVER

The journey begins by laying a foundation. Part 1 introduces the principles of quantum computing, explaining how it differs from classical computing and exploring its historical roots. You'll learn about key concepts like qubits, superposition, and entanglement—the building blocks of quantum computers. Part 2 explores how quantum computing revolutionizes fields such as artificial intelligence, cryptography, and materials science.

In Part 3, we examine its impact on the finance, logistics, and healthcare industries, providing real-world examples and case studies of businesses that leverage quantum technology. Part 4 brings quantum computing closer to everyday life, discussing its potential in energy solutions, smart cities, and personalized medicine. Finally, Part 5 examines the ethical and societal implications of this disruptive technology and offers practical guidance for individuals and organizations to prepare for the quantum future.

WHY QUANTUM COMPUTING MATTERS

Quantum computing matters because it addresses some of humanity's most pressing challenges. Traditional computers struggle with problems involving massive amounts of data or complex interactions, such as simulating molecules for drug discovery or optimizing renewable energy grids. Quantum computers excel in these areas, enabling breakthroughs that could save lives, protect the environment, and drive economic growth. Moreover, the quantum revolution is already underway. Governments and tech giants are investing billions in quantum research, and early-stage quantum computers are beginning to tackle real-world problems. Understanding this technology is not just about staying informed; it's about being prepared for the opportunities and disruptions it will bring.

A TRANSFORMATIVE JOURNEY

As we embark on this journey, remember that quantum computing is not just a technological advancement; it's a paradigm shift. Like the advent of the Internet or the harnessing of electricity, quantum computing has the potential to redefine how we live and work. It challenges us to think differently, innovate, and adapt. Whether you're an optimist envisioning a quantum-powered solution or a skeptic concerned about its implications, one thing is sure: quantum computing will shape the future. Let's explore that future together.

PART 1

UNDERSTANDING QUANTUM COMPUTING

CHAPTER 1

FROM CLASSICAL TO QUANTUM

THE FOUNDATIONS OF CLASSICAL COMPUTING

To understand the significance of quantum computing, we first need to explore its predecessor, classical computing. Classical computers, which have powered technological advancements for decades, rely on binary systems to process information. This binary system uses bits—the most minor units of data—in one of two states: 0 or 1. These bits form the foundation of classical computing, where sequences of 0s and 1s encode instructions and data for processing. By combining bits into more extensive sequences, classical computers execute calculations and solve problems ranging from simple arithmetic to complex simulations, such as weather forecasting or financial modeling.

The power of classical computing lies in its ability to process logical operations systematically. This has enabled revolutions across industries such as communication, where data is transmitted digitally; healthcare, through medical imaging and diagnostics; and finance, with algorithms managing investments and fraud detection. However, classical computing is inherently limited by its sequential nature. Classical systems struggle to scale efficiently when faced with complex problems, such as simulating molecular interactions in drug discovery or analyzing massive datasets for climate prediction. Even with the advent of parallel computing and supercomputers, these systems remain constrained by the binary approach, making specific challenges computationally prohibitive.

As technology evolves, the demand for solving exponentially complex problems has highlighted these limitations. Classical computing processes information step-by-step or within limited parallel frameworks, meaning problems requiring simultaneous evaluation of countless possibilities often exceed its capabilities. This bottleneck has paved the way for the development of quantum computing, a fundamentally different paradigm designed to overcome these challenges and revolutionize our approach to computation.

THE RISE OF QUANTUM COMPUTING

Quantum computing represents a significant leap forward in how computers solve problems. Unlike classical computers, which work with bits that can only be 0 or 1, quantum computers use quantum bits or qubits. Qubits take advantage of quantum mechanics, which is the science that explains how very tiny particles like atoms behave. Two unique features make qubits different from regular bits: superposition and entanglement.

SUPERPOSITION: BEYOND 0 AND 1

Superposition is one of the most fascinating and essential principles of quantum computing. It allows qubits to exist in multiple states simultaneously. In classical computing, a bit can only be in one state at a time—0 or 1. But with superposition, a qubit can simultaneously represent 0, 1, or any combination of these states. This unique property means quantum computers can handle many calculations simultaneously, unlike classical computers, which must process them individually.

Imagine trying to solve a complex puzzle where you must test every possible combination to find the solution. A classical computer would try each combination sequentially, one at a time, until it finds the correct answer. Thanks to superposition, a quantum computer could evaluate many combinations simultaneously, significantly reducing the time required to solve the puzzle.

For instance, superposition can aid in optimization problems, such as determining the most efficient route for delivery trucks in a busy city. A classical computer would analyze each route individually, while a quantum computer could analyze all possible routes simultaneously to determine the best one almost instantly. This ability to parallelize calculations makes superposition such a powerful tool in quantum computing.

However, there is a catch. When you measure a qubit to

observe its state, it collapses into one specific value—0 or 1—, losing its superposition. This makes quantum computing not just about harnessing superposition but also about developing algorithms and techniques that take advantage of it without losing the information stored in these multiple states.

The potential applications of superposition are vast, ranging from faster database searches to advanced simulations of molecular structures for drug discovery. Superposition opens doors to solving problems currently impractical or impossible for classical computers, paving the way for revolutionary advancements in science, technology, and beyond.

ENTANGLEMENT: THE POWER OF CORRELATION

Entanglement is one of the most extraordinary and unique properties of quantum computing. It occurs when two or more qubits become interconnected so that the state of one qubit is directly tied to the state of another, no matter how far apart they are. This phenomenon means that when you measure the state of one qubit, you immediately know the state of its entangled partner, even if vast distances separate them.

This remarkable feature has profound implications for quantum computing. By leveraging entanglement, quantum computers can perform coordinated and highly complex calculations across multiple qubits. Unlike classical computers, which handle data one piece at a time, entangled qubits work in unison, allowing for faster and more efficient problem-solving.

For example, consider a situation where a quantum computer needs to analyze a network of interconnected systems, such as optimizing traffic flow in a busy city. Using entangled qubits, the quantum computer can simultaneously process and correlate data from multiple points, providing insights and solutions that would take classical systems far longer to compute.

Entanglement also plays a crucial role in quantum communication. It enables the development of highly secure communication systems where the laws of quantum mechanics protect the exchange of information. If an entangled particle is tampered with during transmission, the change is immediately detectable, ensuring unparalleled levels of security.

However, achieving and maintaining entanglement is no small feat. Qubits are extremely sensitive to their environment; any disturbance can break their entangled state. This challenge, decoherence, requires advanced techniques to preserve entanglement for practical use in quantum computers. Researchers are continually exploring ways to enhance the stability and reliability of entanglement, making it a key focus of ongoing quantum research.

Entanglement is a cornerstone of quantum computing, unlocking capabilities that far surpass those of classical systems. Entanglement enhances computational power by enabling qubits to work together in a profoundly interconnected way and opens new frontiers in secure communication and data processing. This powerful phenomenon continues to drive the development of quantum technologies, bringing us closer to a future where entanglement transforms how we compute, communicate, and solve some of the world's most complex problems.

THE QUANTUM LEAP: WHY IT MATTERS

The combination of superposition and entanglement gives quantum computers their extraordinary potential. They can tackle problems practically impossible for classical computers to solve within a reasonable timeframe. For instance, quantum computing can revolutionize fields like cryptography by breaking encryption methods that are currently considered secure. It can also accelerate drug discovery by simulating molecular interactions at an unprecedented scale.

CHALLENGES OF QUANTUM COMPUTING

Despite its promise, quantum computing is not without challenges. One major hurdle is maintaining qubit stability. Qubits are highly sensitive to their environment, and any interference can cause them to lose their quantum state—a phenomenon known as decoherence. Additionally, quantum systems require sophisticated error correction techniques to ensure reliable results. Building scalable quantum hardware and algorithms remains an ongoing area of research, with significant progress made by companies such as IBM and Google, as well as startups specializing in quantum technology.

COMPARING CLASSICAL AND QUANTUM COMPUTING

To illustrate the differences between classical and quantum computing, consider a maze-solving scenario. A classical computer navigates the maze by trying each path one by one until it finds the exit. In contrast, a quantum computer explores all paths simultaneously, leveraging superposition to evaluate multiple possibilities simultaneously. This difference highlights why quantum computing is promising for solving problems involving optimization, pattern recognition, and large-scale data analysis.

REAL-WORLD APPLICATIONS OF QUANTUM CONCEPTS

While quantum computing remains in its early stages, its foundational principles are already being applied in areas like quantum cryptography, where qubits are used to create secure communication channels. Another example is quantum sensing, which leverages quantum mechanics to develop exact measurement tools for medical imaging, environmental monitoring, and navigation.

As we transition from classical to quantum computing, we must view these technologies as complementary rather than competitive. Classical computing, with its binary system and established algorithms, will continue to be the backbone of most day-to-day technological applications. Tasks such as managing databases, running enterprise software, and handling routine computational tasks will remain well within the domain of classical computers. Their efficiency, reliability, and scalability make them indispensable for countless industries.

Quantum computing, on the other hand, represents a specialized and transformative addition to our computational toolkit. It expands the boundaries of what is possible, addressing challenges that classical systems cannot efficiently solve. Quantum systems excel in scenarios that involve vast amounts of data, intricate interactions, or the need to evaluate numerous possibilities simultaneously. For example, quantum computing holds unparalleled potential in optimizing complex supply chains, simulating molecular behavior for drug discovery, or breaking down encryption algorithms.

The relationship between classical and quantum computing can be likened to a partnership. Rather than replacing classical systems, quantum computing will augment them, providing enhanced capabilities for specific problems. This complementary dynamic ensures that both technologies will evolve together, strengthening the other's impact.

As businesses, governments, and researchers explore the quantum frontier, collaboration across disciplines will be key. By integrating quantum systems into existing computational frameworks, we can harness the strengths of both technologies to tackle challenges previously thought insurmountable. In this way, the future of computing is not about choosing one over the other but embracing the synergy between classical and quantum paradigms.

This chapter sets the stage for a deeper exploration of the mechanics, history, and transformative potential of quantum computing. By understanding the foundational differences between classical and quantum systems, we can better appreciate their revolutionary implications.

CHAPTER 2

A BRIEF HISTORY OF QUANTUM COMPUTING

THE FOUNDATIONS OF QUANTUM MECHANICS

The journey to quantum computing begins with the emergence of quantum mechanics, a revolutionary field of physics that was developed to explain phenomena that classical mechanics could not. By the late 19th century, scientists had encountered challenges in understanding the behavior of light and matter at microscopic scales. Classical physics, grounded in Newtonian mechanics and Maxwell's electromagnetic theory, was unable to explain specific experimental observations, such as blackbody radiation and the photoelectric effect. These puzzles laid the groundwork for a new paradigm in physics.

One of the first cracks in classical theory came with the problem of blackbody radiation. A blackbody, an idealized object that absorbs and emits all radiation, did not behave as classical theories predicted when heated. Instead of radiating energy uniformly across all wavelengths, the energy peaked at a specific wavelength and decreased at higher frequencies. This discrepancy, known as the ultraviolet catastrophe, puzzled scientists for years.

Max Planck, a German physicist, resolved the issue in 1900 by proposing that energy is not continuous but quantized. He introduced the idea that energy is emitted or absorbed in discrete packets, which he referred to as quanta. This groundbreaking proposal marked the birth of quantum theory and introduced Planck's constant, a fundamental constant that defines the size of these energy packets. Planck's work was initially met with skepticism but proved instrumental in advancing the understanding of atomic and subatomic processes.

Another critical phenomenon that classical physics could not explain was the photoelectric effect, observed when light strikes a metal surface and ejects electrons. According to classical wave theory, increasing light intensity should increase the energy of ejected electrons. However, experiments revealed that the energy of the electrons depended on the light's frequency, not its intensity. This contradiction challenged classical theories and required a new explanation.

Albert Einstein provided the solution in 1905 by extending Planck's quantum hypothesis to light itself. He proposed that light consists of particles, or photons, each carrying a quantized amount of energy proportional to its frequency. This insight not only explained the photoelectric effect but also cemented the concept of wave-particle duality, where light exhibits both particle-like and wave-like behavior. Einstein's work earned him the Nobel Prize in Physics in 1921, setting the stage for further exploration of quantum phenomena.

As quantum mechanics evolved, scientists sought to describe the behavior of particles at the atomic and subatomic levels. Erwin Schrödinger, an Austrian physicist, developed a mathematical framework for this purpose. In 1926, he introduced the Schrödinger equation, a partial differential equation that describes how the quantum state of a physical system changes over time. Schrödinger's equation provided a method for calculating the probability of finding a particle in a particular state or location, a concept central to quantum mechanics.

Schrödinger also illustrated the counterintuitive nature of quantum mechanics with his famous thought experiment, Schrödinger's cat. In this scenario, a cat is placed in a sealed box with a mechanism that has a 50% chance of killing it based on a quantum event. Until the box is opened and observed, the cat exists in a superposition of being alive and dead. This paradox highlights the role of observation in determining the state of a quantum system and underscores the strange and probabilistic nature of quantum mechanics.

QUANTUM THEORY MILESTONES: PLANCK, EINSTEIN, SCHRÖDINGER

The development of quantum mechanics involved a series of groundbreaking contributions that redefined our understanding of the universe's fundamental laws. Max Planck, often regarded as the father of quantum theory, made a pivotal discovery in 1900 while investigating blackbody radiation—the emission of light and heat by an idealized object. Classical theories predicted an infinite amount of energy at short wavelengths, a paradox known as the ultraviolet catastrophe. Planck resolved this by proposing that energy is quantized and emitted in discrete packets called quanta. This radical idea introduced Planck's constant, a fundamental parameter in physics, and signaled the birth of quantum mechanics. Planck's work challenged existing notions of continuity in energy emission, emphasizing the granular nature of energy at the atomic level. This concept addressed longstanding anomalies and inspired a paradigm shift in physics.

Albert Einstein expanded on Planck's work in 1905, addressing another phenomenon that puzzled classical physics: the photoelectric effect. When light strikes certain materials, it ejects electrons, but classical wave theory could not explain why the ejected electrons' energy depended on light's frequency rather than its intensity. Einstein proposed that light itself is quantized, consisting of particles called photons. This groundbreaking insight resolved the photoelectric effect and reinforced the concept of wave-particle duality, a cornerstone of quantum theory. Wave-particle duality suggests that particles, such as photons and electrons, exhibit both particle-like and wave-like behavior, depending on how they are observed. Einstein's contribution earned him the 1921 Nobel Prize in Physics and laid the groundwork for quantum mechanics to challenge longstanding classical assumptions. Moreover, his theories provided a deeper understanding of phenomena such as light's interaction with matter, paving the way for innovations in quantum technology.

Erwin Schrödinger, an Austrian physicist, further advanced quantum mechanics by developing a mathematical framework to describe the behavior of atoms and subatomic particles. In 1926, he introduced the Schrödinger equation, a wave equation that models how quantum states evolve. This equation provided a way to predict the likelihood of finding particles in specific states or locations, offering a probabilistic view of the quantum world. Schrödinger's contributions extended beyond mathematics; his famous thought experiment, Schrödinger's Cat, illustrated the bizarre implications of quantum superposition. In this hypothetical scenario, a cat sealed in a box with a quantum-based mechanism exists simultaneously in both alive and dead states until it is observed. This paradox encapsulates the role of measurement in determining quantum outcomes, highlighting the counterintuitive and probabilistic nature of quantum systems. Schrödinger's work also helped formalize the principle of superposition, where particles can exist in multiple states simultaneously, a phenomenon crucial to quantum computation.

These milestones—Planck's quantization of energy, Einstein's photon theory of light, and Schrödinger's mathematical descriptions—formed the theoretical backbone of quantum mechanics. Each contribution builds upon the last, setting the stage for the eventual development of quantum computing, which leverages these principles to process information in fundamentally new ways. This cumulative progress reflects not only the brilliance of these physicists but also the collaborative nature of scientific discovery as their work continues to inspire and underpin advancements in quantum technology.

EVOLUTION OF QUANTUM COMPUTING: FEYNMAN TO THE PRESENT

The concept of quantum computing emerged from the visionary ideas of physicist Richard Feynman in the early 1980s. Feynman, known for making complex concepts accessible, realized that classical computers struggle to simulate quantum systems due to their complexity and scale. In a 1981 lecture, he proposed using quantum systems as computational tools, harnessing their unique properties to solve problems classical systems couldn't handle. This insight laid the groundwork for the development of quantum computing.

At the same time, David Deutsch reached another significant milestone, formalizing the idea of a universal quantum computer in 1985. Deutsch introduced a theoretical model for computations using quantum bits, or qubits, which utilize quantum phenomena like superposition and entanglement. This model suggested that a quantum computer could outperform classical computers in solving specific problems, especially those requiring complex parallel processing.

By the late 20th century, quantum computing theory advanced further. In 1994, mathematician Peter Shor created an algorithm for factoring large numbers using a quantum computer. This breakthrough demonstrated the potential of quantum computing to disrupt encryption methods like RSA, sparking excitement and concern. Around the same time, Lov Grover introduced an algorithm for database searching, demonstrating the ability of quantum systems to speed up traditionally slow and resource-intensive tasks.

However, practical implementation lagged behind theoretical progress. Creating stable qubits and reducing errors caused by quantum decoherence were significant hurdles. Researchers experimented with technologies such as superconducting circuits, trapped ions, and photonic systems to build reliable quantum computers. By the early 2000s, small quantum devices capable of running basic algorithms emerged, marking the shift from theory to experimentation.

Quantum computing gained momentum in the 2010s, with substantial investments from governments, academia, and private companies. Industry leaders like IBM, Google, and Microsoft spearheaded efforts to create scalable quantum systems. IBM's Quantum Experience, launched in 2016, provided cloud-based access to quantum computing, enabling researchers worldwide to experiment and innovate.

Google made headlines in 2019 with its Sycamore processor, claiming quantum supremacy by performing a task faster than the best classical supercomputers. While this claim sparked debates, it highlighted the rapid progress being made in the field of research.

Innovative startups also contributed significantly. Companies such as Rigetti Computing, IonQ, and D-Wave explored diverse approaches to building quantum systems, ranging from superconducting qubits to quantum annealers. These startups complemented the work of established corporations, creating a dynamic ecosystem of research and development.

Today, the field is advancing remarkably, though challenges remain. Scalable, fault-tolerant quantum computers are still in development, but hybrid systems combining classical and quantum technologies are already showing promise. Governments in the United States, China, and the European Union have launched significant initiatives to lead the quantum revolution.

As quantum computing evolves, breakthroughs in hardware, software, and error correction bring us closer to realizing its full potential. From tackling scientific problems once thought unsolvable to transforming industries such as finance and healthcare, the journey from Feynman's initial idea to today's expanding quantum ecosystem showcases human ingenuity and the relentless drive to innovate.

CURRENT LEADERS: IBM, GOOGLE, MICROSOFT, AND INNOVATIVE STARTUPS

The quantum computing race has drawn some of the world's most influential companies. Each contributes to the field's rapid evolution with unique approaches and technologies. These organizations are pushing the boundaries of what's possible and setting the stage for the widespread adoption of quantum technologies.

IBM is a trailblazer in the quantum computing space, having established itself as a leader with its IBM Quantum initiative. In 2016, IBM launched the IBM Quantum Experience, a cloud-based platform that allowed users worldwide to access and experiment with quantum computers. This democratization of quantum computing spurred innovation across industries and academia. IBM's quantum systems use superconducting qubits, which have shown promising advancements in reliability and scalability. The company's roadmap includes developing fault-tolerant quantum computers capable of solving real-world problems, such as those in materials science, finance, and cryptography.

Google has been a key player, making headlines in 2019 when it claimed quantum supremacy with its Sycamore processor. This achievement demonstrated that a quantum computer could perform a specific calculation faster than the world's most influential classical supercomputers. Google's approach focuses on developing scalable quantum processors while addressing challenges related to error correction and coherence. The company's Quantum AI division actively collaborates with academic and industrial partners to accelerate research and build practical applications for quantum technology.

Microsoft has taken a distinctive approach by focusing on topological qubits, which promise greater stability and lower error rates than other qubit technologies. Microsoft integrates quantum computing with cloud services through its Azure Quantum platform, making it accessible to developers and researchers worldwide. The company also supports quantum programming through tools like Q#, a specialized programming language for quantum algorithms. Microsoft's commitment to building a comprehensive quantum ecosystem underscores its vision for seamlessly integrating quantum computing into existing technologies.

While tech giants dominate headlines, startups have brought innovation and diversity to the quantum computing landscape. Companies like Rigetti Computing are developing full-stack quantum solutions that combine hardware and software to deliver accessible quantum systems. IonQ has pioneered trapped-ion technology for qubits, which offers high fidelity and long coherence times. Their advancements have positioned them as leaders in the commercial quantum space. D-Wave, another prominent startup, specializes in quantum annealing, a niche approach well-suited for optimization problems in logistics, finance, and machine learning.

The global quantum ecosystem also benefits from the efforts of smaller but highly innovative companies, such as Zapata Computing and Cambridge Quantum. These firms focus on quantum software, algorithms, and applications, bridging the gap between hardware capabilities and practical use cases. Their contributions are crucial for realizing the full potential of quantum computing in solving complex problems across various industries.

Collaboration among these leaders is shaping the future of quantum computing. Partnerships between academic institutions, government agencies, and private enterprises foster innovation and accelerate progress. As the competition intensifies, the collective advancements by IBM, Google, Microsoft, and numerous startups underscore the transformative potential of quantum computing.

CHAPTER 3

HOW QUANTUM COMPUTERS WORK

To truly appreciate the transformative potential of quantum computing, it is essential to understand its core mechanisms and operational principles. Quantum computers operate fundamentally differently from classical computers, leveraging the unique properties of quantum mechanics to process information. This chapter will guide you through the cutting-edge hardware technologies that enable these machines, including superconducting qubits, trapped ions, and photonic systems. Additionally, we will explore groundbreaking algorithms such as Shor's algorithm and Grover's search, which exemplify the computational advantages of quantum systems. Finally, the chapter examines the critical challenges of noise, decoherence, and error correction that researchers must address to make quantum computing more robust and widely applicable.

HARDWARE TECHNOLOGIES

Understanding the physical mechanisms that enable quantum computing provides insights into the challenges and possibilities of this transformative technology. Quantum computers process information differently from classical computers by utilizing qubits, which harness quantum mechanical properties like superposition and entanglement. Here, we explore the primary hardware technologies that bring quantum computing to life: superconducting qubits, trapped ions, and photonic systems.

Qubits are the fundamental units of quantum information at the heart of quantum computing. Unlike classical bits, which can only exist in states of 0 or 1, qubits leverage the principles of quantum mechanics to exist in multiple states simultaneously, a property known as superposition. This ability enables quantum computers to perform complex calculations exponentially faster than classical computers for specific problems. However, creating and maintaining qubits is an extraordinary scientific and engineering challenge. Three leading technologies are at the forefront of this endeavor: superconducting qubits, trapped ions, and photonic systems.

SUPERCONDUCTING QUBITS

Superconducting qubits represent a cornerstone of quantum hardware development and are one of the most mature technologies in the field. These qubits operate using superconducting circuits fabricated from materials with zero electrical resistance when cooled to cryogenic temperatures, often achieved using liquid helium. This superconducting state enables the creation of tiny loops of current that form the basis of qubit operation.

Superconducting qubits are today's most advanced and widely used quantum computing technologies. In a superconducting qubit system, Josephson junctions—specialized circuit elements—allow precise control over the quantum states by facilitating the tunneling of electron pairs. These circuits interact with microwave photons to create and manipulate quantum states, making them ideal for fast and highly responsive quantum operations.

Companies like IBM and Google rely heavily on superconducting qubits due to their scalability and relatively fast gate operation times. For example, Google's Sycamore processor, which achieved quantum supremacy in 2019, utilizes superconducting qubits. These circuits are coupled with microwave resonators to manipulate and read qubit states, enabling complex computations. Despite their promise, superconducting qubits face challenges like noise and decoherence, which we will discuss later.

TRAPPED IONS

Trapped ion systems take an entirely different approach to quantum computation by using individual atomic ions as qubits. These ions are held in place by sophisticated electromagnetic traps generated within ultra-high vacuum chambers to eliminate external interferences. The ions' quantum states are manipulated using precisely tuned lasers interacting with their internal energy levels. Trapped ion systems represent another leading approach to building quantum computers. In this method, individual ions are confined and suspended in electromagnetic fields within ultra-high vacuum chambers. Each ion represents a qubit, with its electronic states encoding quantum information. By applying laser pulses of varying intensities and durations, researchers can perform quantum gates and create entangled states, which are essential for quantum computations. The stability and longevity of trapped ion qubits make them highly reliable for executing algorithms that demand precision.

Trapped ion systems excel in precision and coherence times, often outperforming superconducting qubits in terms of instability. IonQ and Honeywell are notable companies pioneering this technology. However, scalability remains a challenge due to the complexity of precisely controlling large numbers of ions. The slower gate operation times compared to superconducting qubits are another area for improvement.

PHOTONIC SYSTEMS

Photonic quantum computers leverage light particles, or photons, to encode and process quantum information. This approach has unique advantages, particularly in quantum communication and distributed quantum computing, where photons' natural resilience to noise plays a critical role.

Photonic systems leverage particles of light (photons) to encode and process quantum information. As carriers of quantum information, photons are inherently less susceptible to environmental interference than other qubit systems. Quantum information is encoded in properties such as polarization, phase, or time-bin states, which can be manipulated using advanced optical components like beam splitters, interferometers, and photodetectors. Photonic quantum computers utilize specialized devices, such as beam splitters, waveguides, and detectors, to control and measure photon states.

Xanadu, a leading company in this field, is developing photonic quantum computers that can solve complex optimization problems. While photonic systems show great promise for specific applications, their practical implementation for general-purpose quantum computing is still in its infancy.

QUANTUM ALGORITHMS

Quantum algorithms are the computational blueprints that allow quantum computers to outperform classical systems for specific tasks. These algorithms offer unprecedented efficiency in solving specific problems by leveraging principles like superposition and entanglement. This section explores two groundbreaking algorithms that highlight the capabilities of quantum computing: Shor's algorithm and Grover's search algorithm.

The power of quantum computing lies in its hardware and unique algorithms, which exploit quantum phenomena to solve problems more efficiently than classical algorithms. Two of the most well-known quantum algorithms are Shor's and Grover's search algorithms.

SHOR'S ALGORITHM

Shor's algorithm, developed by mathematician Peter Shor in 1994, stands as one of the most transformative breakthroughs in quantum computing. Its primary function is to factorize large integers exponentially faster than any classical algorithm. The implications of this capability are profound, particularly in the realm of cryptography. Modern encryption systems, such as RSA, depend on the difficulty of factoring large composite numbers into their prime components. Classical computers require billions of years to factorize such numbers, but Shor's algorithm can achieve this task in mere hours or even minutes, given a sufficiently powerful quantum computer.

Developed by Peter Shor in 1994, Shor's algorithm is designed to factorize large integers exponentially faster than the best-known classical algorithms. This breakthrough poses a direct challenge to traditional cryptographic protocols. For example, RSA encryption—widely used to secure online communications, financial transactions, and governmental data—relies on the impossibility of factoring in large numbers. Shor's algorithm could render these systems obsolete, sparking a global shift towards quantum-safe encryption methods. This potential impact underscores the urgency of advancing quantum computing hardware and cryptographic innovation to mitigate future vulnerabilities. A sufficiently powerful quantum computer running Shor's algorithm could render much of today's encryption obsolete, necessitating the development of quantum-safe cryptographic protocols.

GROVER'S SEARCH ALGORITHM

Developed by Lov Grover in 1996, Grover's algorithm is another landmark in quantum computing. Unlike Shor's algorithm, which offers exponential speedup, Grover's algorithm provides a quadratic improvement for unstructured search problems. It can solve tasks like database searches or optimization problems significantly faster than its classical counterparts.

Grover's algorithm, developed by Lov Grover in 1996, provides a quadratic speedup for unstructured search problems. Grover's algorithm is particularly effective for tasks that require identifying a solution within a large, unsorted dataset. For instance, a classical computer would require N steps to guarantee success if a database contains N entries and a specific item needs to be found. Grover's algorithm reduces this to approximately sqrt(N) steps, providing a substantial efficiency boost. This quadratic speedup has practical applications in fields like logistics, cryptography, and artificial intelligence, where search problems are prevalent. For example, if a classical computer requires steps to find a specific item in an unsorted database, Grover's algorithm can achieve the same result in approximately steps.

These algorithms highlight the unique strengths of quantum computing but underscore the need for scalable, error-resistant quantum hardware to realize its full potential.

CHALLENGES

While quantum computing holds tremendous promise, it is far from realizing its full potential due to significant technical challenges. These obstacles stem from the inherent fragility of quantum systems, which are highly susceptible to environmental and operational disturbances. The ability to harness and stabilize qubits in practical, large-scale systems is a monumental task. Three primary challenges—noise, decoherence, and the implementation of robust error correction—stand as barriers that researchers and engineers are striving to overcome. These challenges are not just technical hurdles but also define the roadmap for scaling quantum technologies to real-world applications, making them some of the most critical areas of focus in quantum computing research.

NOISE

Noise refers to any unwanted disturbances that disrupt the quantum states of qubits. In quantum systems, these disturbances can stem from environmental factors such as temperature fluctuations, electromagnetic interference, and even cosmic radiation. Noise introduces errors in computations by causing qubits to deviate from their intended states, undermining the reliability of quantum operations. Managing noise is a formidable challenge because quantum systems require extreme isolation to maintain their coherence. Researchers employ techniques such as cryogenic cooling and advanced shielding to minimize these disturbances, but achieving the necessary isolation level often results in complex and expensive setups. This sensitivity makes it challenging to perform reliable quantum computations over extended periods of time.

DECOHERENCE

Decoherence occurs when qubits lose their quantum state due to interactions with their surroundings. This process effectively erases the quantum information stored in a system, limiting the time available for computations. Decoherence is one of the primary reasons quantum computers struggle to perform extended calculations without errors. The coherence time of a quantum system is a critical parameter that determines how long qubits can maintain their superposition and entanglement. Short coherence times hinder the execution of complex algorithms as errors accumulate rapidly. Advances in materials science and qubit design are being explored to extend coherence times and enhance system reliability. Improving coherence times is essential for executing more complex quantum algorithms without errors.

ERROR CORRECTION

Error correction in quantum computing is far more complex than in classical systems due to the unique nature of quantum errors. Qubits are susceptible to bit-flip and phase-flip errors, unlike classical bits, which can experience simple bit-flip errors. Moreover, correcting these errors must be done without directly observing the quantum states, as measurement collapses the superposition. Classical computers rely on error-correcting codes to ensure reliable computations, but quantum error correction is far more complex. To address this, researchers have developed quantum error-correcting codes, such as the surface and topological codes. These methods involve encoding logical qubits into multiple physical qubits to detect and correct errors indirectly. However, implementing these codes requires significant overhead, often necessitating hundreds of physical qubits to create a single error-resistant logical qubit. While error correction is essential for building scalable quantum computers, it comes at the cost of substantial resource demands. Efforts are ongoing to optimize these techniques and reduce the overhead, with researchers exploring innovative approaches such as quantum error mitigation and fault-tolerant architectures.

Understanding how quantum computers work is the first step in appreciating their transformative potential. While current systems are far from perfect, advances in hardware technologies, algorithm development, and error correction steadily bring us closer to a quantum future. The following chapters will explore how this revolutionary technology is already beginning to reshape industries and solve problems that were once thought unsolvable.

PART 2

QUANTUM COMPUTING'S IMPACT ON TECHNOLOGY

CHAPTER 4

REVOLUTIONIZING ARTIFICIAL INTELLIGENCE AND MACHINE LEARNING

Artificial Intelligence (AI) and Machine Learning (ML) have become transformative forces across industries, revolutionizing tasks from image recognition to natural language processing (NLP). Yet, as these technologies evolve, they face limitations in computational power and efficiency. Quantum computing offers a solution that promises to accelerate AI and ML tasks, unlocking capabilities previously inaccessible to classical computers.

The rapid growth of AI and machine learning (ML) has reshaped how humans interact with technology. These disciplines have enabled breakthroughs such as virtual assistants that can understand natural speech, self-driving cars that process real-time data to make decisions, and algorithms that detect fraudulent transactions within milliseconds. Despite their widespread success, these technologies are fundamentally constrained by the capabilities of classical computing systems.

Classical computers, the backbone of AI and ML today, rely on binary processing units to execute operations. While advancements in classical hardware, including GPUs and TPUs, have driven progress, they are reaching their practical limits in handling the immense computational demands of modern AI systems. For example, training a state-of-the-art deep learning model can require weeks or months on even the most advanced supercomputers, consuming vast energy and resources. This creates bottlenecks in innovation and deployment, particularly in applications that require real-time decision-making or the analysis of large datasets.

Consider the example of image recognition. AI systems used for medical imaging, such as those that detect tumors in CT scans, must process thousands of high-resolution images. The accuracy of these systems depends on their ability to extract subtle patterns from data, a task that often requires iterative fine-tuning of complex algorithms. As datasets grow and algorithms become more sophisticated, the computational burden increases exponentially.

Similarly, natural language processing has seen a surge in complexity. From translating entire books in seconds to understanding human emotions in chat interactions, NLP models like OpenAI's GPT or Google's BERT are expected to perform at near-human levels. However, these models involve billions of parameters and require immense computational power to train and deploy. This makes scaling NLP systems for broader and more nuanced applications challenging and resource-intensive.

Optimization, another critical area within AI and ML, involves finding the optimal solution among numerous alternatives. Whether tuning hyperparameters for a machine learning model or determining the optimal logistics for supply chains, these problems often involve navigating astronomical possibilities. Classical algorithms handle such tasks sequentially, which can be inefficient and time-consuming, especially for highly complex scenarios.

Quantum computing emerges as a game-changing force in this context. By leveraging principles of quantum mechanics, quantum computers introduce a novel approach to information processing that can overcome the limitations of classical systems. With their unparalleled ability to handle vast amounts of data and explore multiple possibilities simultaneously, quantum computers promise to revolutionize AI and ML. As industries increasingly rely on these technologies to solve real-world problems, the integration of quantum computing offers a pathway to faster solutions and entirely new capabilities that classical computers cannot achieve.

QUANTUM COMPUTING'S ROLE IN AI AND ML

Quantum computing's potential lies in its ability to process and analyze data at unprecedented speeds. Unlike classical computers, which operate in binary, quantum computers leverage the principles of quantum mechanics—superposition and entanglement—to perform parallel computations. This quantum advantage enables exponentially faster calculations for specific problems, making it particularly valuable in AI and ML applications, where processing massive datasets and solving complex mathematical problems are routine challenges.

In Machine Learning, one of the most time-intensive tasks is training models, which involves adjusting billions of parameters across multiple iterations to minimize errors. Quantum computers can optimize this process by exploring multiple parameter configurations simultaneously. This quantum parallelism significantly reduces training times, allowing for the creation of more accurate models in less time. Techniques such as Quantum Support Vector Machines (QSVMs) and Variational Quantum Classifiers (VQCs) exemplify how quantum resources can be harnessed to enhance learning algorithms.

Another critical contribution of quantum computing is improving optimization tasks foundational to AI. Traditional optimization methods often involve iterative searches that are computationally expensive. Quantum algorithms, such as Grover's search and Quantum Approximate Optimization Algorithms (QAOAs), accelerate these processes by evaluating potential solutions concurrently, achieving superior results with far fewer iterations. This capability improves efficiency and unlocks new potential for solving previously considered intractable problems.

In AI decision-making processes, quantum-enhanced algorithms add layers of sophistication. By leveraging quantum entanglement, systems can more accurately analyze interdependent variables in complex datasets, producing insights that guide smarter decisions. For example, applications in predictive analytics benefit from this capacity to evaluate numerous potential scenarios simultaneously, leading to optimized outcomes with reduced uncertainty

QUANTUM ACCELERATION OF AI AND ML TASKS

One of the most transformative aspects of quantum computing is its ability to accelerate AI and ML tasks drastically. Traditional AI and ML models rely heavily on iterative optimization and vast computational resources to train algorithms, identify patterns, and produce predictions. Quantum computing revolutionizes this by introducing parallelism on a scale unmatched by classical systems.

A key challenge in AI and ML is training models, which often involves billions of parameters that require fine-tuning over multiple epochs. Quantum computers can significantly shorten this process through quantum parallelism, where multiple potential solutions are evaluated simultaneously. For instance, variational quantum algorithms are particularly suited to optimizing neural networks by leveraging quantum states to explore complex solution spaces more efficiently.

Another critical area is feature selection and dimensionality reduction. Due to their size and complexity, high-dimensional datasets — such as those used in genomics or image processing, pose challenges for classical systems. Quantum computing excels here, using algorithms like the Quantum Approximate Optimization Algorithm (QAOA) to identify the most relevant features and discard noise. This capability accelerates computations and improves the overall performance of AI models.

Furthermore, quantum computing enables the development of enhanced generative models, such as those used in creating synthetic data or simulating real-world scenarios. Quantum-enhanced generative adversarial networks (QGANs) can model complex distributions far more effectively than classical GANs, making them invaluable in applications ranging from drug discovery to creative industries.

In reinforcement learning, a subfield of AI that focuses on decision-making and strategy optimization, quantum computing offers groundbreaking advancements. Classical reinforcement learning relies on trial and error to evaluate different strategies. Quantum reinforcement learning can process multiple strategies simultaneously, rapidly identifying the optimal path in complex environments such as autonomous driving or dynamic resource allocation.

Optimization tasks greatly benefit from quantum speedups, a cornerstone of AI and machine learning. Algorithms like Grover's search provide quadratic improvements in unstructured search problems. At the same time, quantum-inspired techniques solve highly constrained problems, such as vehicle routing or production scheduling, with unprecedented efficiency. These advancements are crucial for real-time AI applications that demand swift decision-making.

Quantum computing enhances the interpretability of AI. Understanding why AI models make specific predictions is a growing area of focus. Quantum techniques can uncover hidden relationships in data by analyzing quantum states, providing new insights into the decision-making processes of AI systems. This enhances trust in AI and opens doors to more ethical and accountable applications.

APPLICATIONS OF QUANTUM COMPUTING IN AI AND ML

1. IMAGE RECOGNITION

One of the most impactful applications of AI is image recognition, which is used in various fields, including healthcare and security. Classical image recognition algorithms often struggle with large datasets, requiring extensive processing time to identify patterns and anomalies. Quantum computing can accelerate these processes by enhancing feature extraction and classification tasks. For example, quantum-enhanced neural networks can process images more efficiently, enabling faster and more accurate identification of objects, faces, and medical conditions such as tumors in radiology scans.

2. NATURAL LANGUAGE PROCESSING (NLP)

Natural Language Processing powers virtual assistants, chatbots, and translation tools. Current NLP systems rely on deep learning algorithms that require vast computational resources. Quantum computing can streamline these processes by optimizing language models and enabling faster training on large text corpora. This enhancement increases the efficiency of NLP systems and improves their accuracy, making interactions more natural and context-aware.

Consider machine translation as an example. Traditional systems often struggle to capture nuances such as idioms and cultural context. Quantum-enhanced NLP could analyze vast linguistic datasets simultaneously, improving the precision and fluency of translations.

3. Optimization Problems

Optimization is at the heart of many AI and ML tasks, from designing efficient supply chains to fine-tuning hyperparameters in machine learning models. Classical algorithms often rely on iterative approaches, which can be time-consuming and computationally expensive. Quantum computing offers a parallelized approach, enabling faster convergence to optimal solutions.

One notable example is reinforcement learning, a branch of artificial intelligence (AI) focused on decision-making. Quantum algorithms can solve reinforcement learning problems more efficiently by evaluating multiple scenarios simultaneously, allowing AI systems to learn strategies for complex tasks such as autonomous vehicle navigation or dynamic pricing in e-commerce.

CHALLENGES

While the potential of quantum computing in AI and machine learning (ML) is immense, challenges remain. Quantum computers are still in their early stages, with limited qubit counts and issues such as noise and decoherence affecting their performance. Researchers are actively working on error correction methods and hybrid models that combine quantum and classical computing to overcome these limitations. Developing quantum algorithms for AI and machine learning (ML) requires expertise in both domains. This interdisciplinary challenge has spurred collaborations between tech giants, universities, and startups, driving innovation in the field.

Integrating quantum computing with AI and ML marks the beginning of a new era in technology. As quantum hardware matures and algorithms improve, we can expect breakthroughs in areas like personalized medicine, climate modeling, and financial analysis. These advancements will enhance existing AI applications and create new possibilities, from predicting complex biological systems to designing next-generation materials.

Quantum computing's ability to accelerate AI and ML tasks represents a paradigm shift in processing and analyzing information. By addressing computational bottlenecks and enabling innovative solutions, quantum technology is poised to redefine the future of artificial intelligence, making it more powerful and pervasive than ever before.

CHAPTER 5

QUANTUM IN CRYPTOGRAPHY AND CYBERSECURITY

In an era where digital transactions, secure communications, and online data exchanges underpin nearly every aspect of modern life, cryptography is the invisible shield safeguarding sensitive information from unauthorized access. Cryptographic protocols such as RSA (Rivest-Shamir-Adleman) and ECC (Elliptic Curve Cryptography) rely on the computational infeasibility of some mathematical issues, ensuring the security of billions of financial transactions, private communications, and military secrets worldwide. However, the advent of quantum computing presents a paradigm shift that threatens to dismantle this foundational security framework.

Quantum computers, unlike their classical counterparts, harness the principles of quantum mechanics—superposition, entanglement, and quantum interference—to process information in fundamentally new ways. Algorithms like Shor's algorithm harness this power to solve problems that classical computers find intractable, doing so efficiently. For RSA, this means factoring large integers into primes at unprecedented speeds. For ECC, it implies solving discrete logarithmic problems in elliptic curve groups relatively efficiently. These breakthroughs, once theoretical, are now within the realm of possibility as quantum technology progresses.

This chapter dives into the profound implications of quantum computing for cryptography and cybersecurity. We will explore how these quantum threats undermine the foundations of current encryption methods, examine the urgent need for post-quantum cryptographic solutions, and shed light on the race to develop protocols that can withstand the quantum era. The urgency is apparent: the quantum revolution is not a distant future but a fast-approaching reality, demanding proactive measures to secure tomorrow's digital infrastructure.

QUANTUM THREATS TO RSA AND ECC ENCRYPTION

Traditional encryption techniques, such as RSA and ECC, rely on computational infeasibility to solve some mathematical issues, which form the backbone of digital security today. RSA, for instance, is based on the challenge of factoring large composite numbers into their prime components. This task requires exponentially increasing time and resources as the numbers grow. Similarly,

ECC depends on the difficulty of solving the discrete logarithm problem in elliptic curve groups, which involves finding the integer multiplier that relates two points on a curve. These problems are computationally expensive for classical computers, making them practically unsolvable within the lifetime of current technology and ensuring that encrypted data remains secure for decades. However, this strength against classical computational methods becomes a vulnerability in the face of quantum computing, which fundamentally alters the rules of computational efficiency. Quantum computing has changed this paradigm. Shor's groundbreaking quantum algorithm can solve these problems exponentially faster than any classical algorithm. A sufficiently powerful quantum computer could factorize the large integers used in RSA encryption or solve the discrete logarithm problems underlying ECC in hours or even minutes. This poses an existential threat to modern encryption standards, which protect everything from financial transactions to military communications.

To fully grasp the seriousness of this threat, it is essential to consider the timeline. Current quantum computers are still in their infancy, lacking the necessary number of stable qubits and the advanced error-correction capabilities required to execute Shor's algorithm effectively on a scale large enough to break modern encryption standards, such as RSA-2048. However, the pace of research and development in quantum computing is accelerating. Leading tech companies, academic institutions, and governments are investing heavily in quantum technologies, with some breakthroughs suggesting that a quantum computer capable of breaking RSA-2048 encryption could be developed within the next two to three decades.

The implications of this timeline are profound. Even though a practical quantum computer does not exist, sensitive data encrypted today could be harvested and stored by malicious actors, only to be decrypted later once quantum capabilities are available. This concept, often referred to as "harvest now, decrypt later," poses significant risks for information that must remain confidential for decades, such as government secrets, intellectual property, and personal health records.

Given the lengthy process required to develop, test, and implement quantum-safe encryption methods, the time to act is now. Transitioning to quantum-resistant cryptography requires years of rigorous research and standardization, as well as widespread adoption and infrastructure upgrades. Organizations must proactively prepare by evaluating their current cryptographic systems, exploring hybrid models that combine classical and quantum-safe methods, and educating stakeholders about the urgency of becoming quantum-ready.

DEVELOPING QUANTUM-SAFE CRYPTOGRAPHY

Recognizing the looming quantum threat, researchers and organizations worldwide are racing to develop quantum-safe cryptographic protocols. These protocols aim to resist classical and quantum attacks, ensuring long-term data security. Post-quantum cryptography (PQC) has emerged as a focal point in this effort.

1. LATTICE-BASED CRYPTOGRAPHY

Lattice-based cryptography is one of the most promising candidates for post-quantum security. It relies on the hardness of problems like the Learning with Errors (LWE) problem, which remains challenging for both classical and quantum computers. Lattice-based schemes are versatile, supporting a wide range of cryptographic applications, including encryption, digital signatures, and key exchange protocols.

2. CODE-BASED CRYPTOGRAPHY

Code-based cryptography, inspired by error-correcting codes, is another robust approach to cryptography. The McEliece cryptosystem, proposed in the 1970s, is an example of a code-based scheme that has withstood decades of cryptanalysis. Its large key sizes, however, pose practical challenges for widespread adoption.

3. MULTIVARIATE CRYPTOGRAPHY

Multivariate cryptography involves solving systems of multivariate quadratic equations, a problem known to be NP-hard. These schemes are particularly suited for digital signature applications but face challenges regarding efficiency and scalability.

4. HASH-BASED CRYPTOGRAPHY

Hash-based cryptographic methods rely on the security of

cryptographic hash functions. These schemes are well-suited for digital signatures and have a straightforward mathematical foundation. The LMS (Leighton-Micali Signature) and XMSS (eXtended Merkle Signature Scheme) are notable examples.

5. ISOGENY-BASED CRYPTOGRAPHY

Isogeny-based cryptography leverages the mathematical properties of elliptic curves to create secure protocols. These schemes are attractive due to their small key sizes, which facilitate efficient implementation.

STANDARDIZATION AND IMPLEMENTATION

The National Institute of Standards and Technology (NIST) spearheads the effort to standardize post-quantum cryptographic algorithms. In 2016, NIST initiated a multi-round competition to evaluate and select quantum-resistant algorithms. The final selection, expected in the mid-2020s, will provide a roadmap for adopting quantum-safe standards.

Transitioning to these standards is a complex process. It involves updating existing cryptographic infrastructure, ensuring interoperability with legacy systems, and educating stakeholders about the importance of quantum readiness. Organizations must adopt a hybrid approach, combining classical and quantum-safe methods during transition.

THE ROLE OF QUANTUM CRYPTOGRAPHY

While post-quantum cryptography focuses on classical algorithms that are resistant to quantum attacks, quantum cryptography explores protocols based on the principles of quantum mechanics. Quantum Key Distribution (QKD) is the most well-known example. QKD enables two parties to share encryption keys with provable security, leveraging the principles of quantum entanglement and the no-cloning theorem. However, QKD faces practical limitations, including high implementation costs and reliance on specialized hardware.

The rise of quantum computing represents both a challenge and an unprecedented opportunity for cryptography and cybersecurity. On one hand, RSA and ECC—the cornerstones of modern cryptographic security—face inevitable obsolescence in the face of quantum advancements. Their mathematical foundations, once unassailable, can be dismantled by quantum algorithms like Shor's in mere moments under sufficient computational power. On the other hand, this emerging technology catalyzes innovation, creating quantum-safe cryptographic methods designed to withstand classical and quantum threats.

By embracing post-quantum cryptography, which leverages robust mathematical frameworks resistant to quantum attacks, and exploring novel quantum cryptographic protocols such as Quantum Key Distribution (QKD), we can construct a digital infrastructure that not only survives but also thrives in the quantum era. This infrastructure will demand continuous collaboration among researchers, industries, and governments to standardize and implement these cutting-edge solutions.

As we navigate this transformative journey, foresight and innovation are paramount. The road ahead is not without its hurdles. Still, with proactive measures, clear strategies, and a commitment to resilience, the digital world can remain secure and adaptable in the face of quantum-powered change.

CHAPTER 6

SIMULATING THE IMPOSSIBLE

QUANTUM SIMULATIONS

The world around us operates on the principles of quantum mechanics, governing everything from the interactions of subatomic particles to the behaviors of complex molecular systems and, more significantly, quantum phenomena. These systems exhibit highly intricate behaviors, dictated by probabilities and quantum states, that are interconnected in ways that defy classical intuition. Modeling such phenomena presents an immense computational challenge due to the exponential growth of variables involved. As the number of particles and their quantum states increases, the complexity of interactions multiplies exponentially, rendering even the most potent classical supercomputers insufficient for accurate simulations. Classical computers, designed to process data sequentially and limited to binary logic, fail to capture the nuanced, probabilistic nature of quantum mechanics. This limitation results in approximations that quickly lose accuracy as the scale and intricacy of the system grow. Consequently, many aspects of quantum behavior have remained elusive, and a computational approach capable of directly handling quantum phenomena is yet to be developed. Quantum computing, however, is specifically designed to address these challenges. By leveraging the fundamental principles of quantum mechanics, such as superposition, entanglement, and quantum interference, quantum computers can directly process and represent complex quantum states. This ability allows them to tackle problems with an exponential number of variables, something classical computers cannot do. For example, a classical computer must approximate the interactions of countless particles and states when simulating quantum systems, which becomes increasingly cumbersome as complexity increases. Quantum computers, on the other hand, can naturally model these interactions due to their inherent quantum nature, making them uniquely suited to explore systems at the atomic and subatomic levels. This capability is not just theoretical; it is already applied in cutting-edge research to unlock new possibilities in materials science, chemistry, biology, and beyond. By embracing this revolutionary tool, scientists are venturing into realms once deemed unattainable, pushing the boundaries of human knowledge and technological advancement.

QUANTUM SIMULATIONS IN MATERIALS SCIENCE

Materials science is one of the most promising areas where quantum simulations are already making a significant impact.

Designing new materials for superconductors, solar cells, or lightweight alloys requires a deep understanding of molecular and atomic interactions. Quantum mechanics governs these interactions and involves a level of complexity that classical computational models often cannot handle due to their limitations in processing exponential variables. Quantum computers, with their ability to process multiple states simultaneously, provide unparalleled precision in modeling these intricate systems.

One groundbreaking area of research involves high-temperature superconductors, materials that could revolutionize energy transmission by eliminating electrical resistance. Quantum simulations enable researchers to investigate the microscopic properties of these materials, revealing specific atomic arrangements and electron interactions that contribute to superconductivity. This insight accelerates the discovery and development of cost-effective and practical superconductors for widespread use, potentially transforming the efficiency of power grids and electronic devices worldwide.

Another transformative application is in the development of advanced solar panel materials. Using quantum simulations, researchers can analyze how various molecular structures interact with photons to absorb and convert sunlight into electricity. This detailed understanding enables the design of more efficient photovoltaic materials, increasing energy conversion rates and reducing costs. These advancements push the boundaries of renewable energy technologies, contributing to a more sustainable future while addressing the growing global demand for clean energy solutions.

REVOLUTIONIZING CHEMISTRY WITH QUANTUM COMPUTING

Chemistry is, by its very nature, governed by the principles of quantum mechanics. Quantum rules dictate the behavior of electrons within molecules, the intricate processes of bond formation and dissociation, and the emergent properties of complex compounds. These highly nonlinear phenomena involve interdependent quantum states in ways that classical computational methods cannot fully replicate. Classical computers quickly reach a computational limit for large molecules with many interacting electrons, necessitating approximations that often compromise accuracy. Quantum computers, however, are uniquely equipped to tackle these challenges. They can model chemical processes with unprecedented precision by directly representing and manipulating quantum states. This capability enables researchers to explore the quantum behavior of molecules, uncovering interactions and dynamics that were previously hidden. For instance, quantum computing enables the detailed study of reaction pathways, transition states, and energy landscapes, which are crucial for understanding and optimizing chemical reactions. Such insights are theoretical yet have practical applications in the development of new drugs, sustainable industrial processes, and innovative materials, ultimately transforming the field of chemistry.

One notable example is the simulation of catalysts, which speed up chemical reactions. Catalysts play a critical role in industries ranging from pharmaceuticals to energy production. Quantum simulations enable researchers to study and design catalysts at the atomic level, optimizing their efficiency and reducing costs. This has profound implications for developing cleaner industrial processes and more sustainable chemical manufacturing.

Additionally, quantum computing is aiding in the study of complex chemical reactions, such as those involved in nitrogen fixation. This natural process, crucial for producing fertilizers, currently requires energy-intensive industrial methods. Quantum simulations could lead to the development of energy-efficient alternatives, thereby reducing the environmental impact of agriculture.

ADVANCES IN DRUG DISCOVERY

The pharmaceutical industry is another field poised to benefit immensely from quantum simulations. Drug discovery involves identifying molecules that interact with biological targets, often requiring years and billions of dollars. Quantum computing has the potential to significantly streamline this process by modeling molecular interactions with unprecedented accuracy.

For instance, quantum simulations can predict how a drug

candidate will bind to a specific protein, a critical step in determining its efficacy and safety. This capability enables researchers to identify the most promising candidates early in development, thereby saving time and resources. Moreover, quantum computers can explore vast chemical spaces to identify novel compounds, opening new avenues for treating untreatable diseases.

In recent years, quantum simulations have been applied to studying Alzheimer's disease, cancer, and antibiotic resistance. Researchers gain valuable insights into potential treatments by simulating the molecular mechanisms underlying these conditions. Quantum computing also facilitates the design of personalized medicines, tailoring treatments to individual patients' unique genetic and molecular profiles.

TRANSFORMING ENERGY STORAGE TECHNOLOGIES

Energy storage is a cornerstone of modern technology, underpinning everything from electric vehicles to renewable energy systems. However, the development of efficient, high-capacity batteries remains a significant challenge. Quantum simulations are invaluable in addressing this issue by providing a detailed understanding of the electrochemical processes within batteries. One of the most significant applications is the design of new battery materials. Quantum computing enables researchers to simulate the interaction of different materials at the atomic level, identifying combinations that maximize energy density and longevity. For example, studies on solid-state batteries—a promising alternative to traditional lithium-ion batteries—have been accelerated by quantum simulations. These batteries offer higher energy storage capacity, improved safety, and longer lifespans, making them ideal for electric vehicles and renewable energy systems.

Another area of impact is the development of hydrogen storage technologies. Hydrogen is a clean energy carrier with immense potential, but storing it efficiently has been a persistent challenge. Quantum simulations enable researchers to understand and optimize materials for hydrogen adsorption and release, thereby paving the way for practical and scalable hydrogen energy solutions.

CHALLENGES AND FUTURE PROSPECTS

While the potential of quantum simulations is immense, the path forward is not without its hurdles. Quantum computers are still in their infancy, operating with few qubits and prone to errors. These errors, caused by noise and decoherence, arise from the fragile quantum states that are easily disrupted by environmental factors such as temperature fluctuations and electromagnetic interference. Overcoming these challenges will require significant advancements in quantum hardware and sophisticated error-correction techniques to stabilize quantum operations and extend their computational power.

Equally critical is the creation of robust quantum algorithms tailored to specific applications. Unlike classical algorithms, which are well-established and highly optimized, quantum algorithms are in their early stages of development. Researchers must design and refine these algorithms to harness the unique capabilities of quantum systems, ensuring they can effectively and efficiently address real-world problems.

Despite these obstacles, the progress made so far is remarkable. Governments worldwide are investing billions of dollars in quantum research initiatives, recognizing its potential to drive economic growth and scientific breakthroughs. Academic institutions are at the forefront of pioneering new quantum theories and practical experiments, while industry leaders such as IBM, Google, and Microsoft are racing to develop scalable quantum systems. The collaboration between these sectors accelerates advancements in hardware and software, laying a robust foundation for the quantum future.

As these efforts continue, the potential applications of quantum simulations in fields like materials science, chemistry, and biology are expanding rapidly. The ability to model complex systems at an atomic level unlocks solutions to challenges once deemed insurmountable. From discovering revolutionary materials to simulating intricate biological processes, quantum computing is reshaping the boundaries of what technology can achieve.

Quantum simulations represent a paradigm shift in our ability to understand and manipulate the natural world. From designing revolutionary materials to advancing drug discovery and energy storage, the impact of this technology is profound and far-reaching. As we continue to push the boundaries of quantum computing, the impossible is becoming possible, unlocking solutions to some of humanity's most pressing challenges.

PART 3

QUANTUM COMPUTING IN BUSINESS

CHAPTER 7

QUANTUM ADVANTAGE IN FINANCE

QUANTUM IN FINANCE

Finance is a domain characterized by intricate interconnections and vast datasets, where activities such as portfolio management, risk analysis, fraud detection, and algorithmic trading require exceptional computational power. Despite their advancements, traditional computers often falter under these burdens, particularly when faced with complex or dynamic problems. Quantum computing emerges as a revolutionary alternative, employing principles of quantum mechanics to solve challenges that classical systems struggle to address. Its ability to simultaneously perform multiple calculations and analyze interconnected variables transforms how financial institutions approach decision-making and problem-solving. This chapter explores the revolutionary potential of quantum computing in transforming finance, with a focus on portfolio optimization, risk analysis, and fraud detection.

PORTFOLIO OPTIMIZATION

Portfolio optimization is one of the most challenging problems in finance, requiring analysts to identify the optimal mix of assets that balances maximum returns with minimal risk. The complexity of this task escalates exponentially as the number of assets increases, introducing a combinatorial explosion of possibilities that even the most advanced classical algorithms struggle to manage effectively. These algorithms typically rely on approximations and heuristic methods, which, while functional, often fall short of delivering genuinely optimal solutions. For example, in large-scale investment portfolios, classical methods might take hours or even days to compute allocations, limiting their ability to adapt to rapidly changing market conditions. The introduction of quantum computing offers a revolutionary approach. Algorithms like the Quantum Approximate Optimization Algorithm (QAOA) can simultaneously evaluate many potential portfolio configurations by harnessing quantum mechanics. This capability expedites the process and enhances the precision of the results, enabling analysts to optimize portfolios in real time as market data evolves. Such advancements promise to redefine how asset management is approached, offering unprecedented accuracy and efficiency in the pursuit of financial growth.

Quantum computing introduces algorithms, such as the Quantum Approximate Optimization Algorithm (QAOA), which excel at tackling complex combinatorial challenges inherent in portfolio optimization. These algorithms leverage the principles of quantum superposition, enabling them to process and evaluate numerous potential portfolio configurations simultaneously, a capability that classical systems can only achieve sequentially and at a significantly slower pace. The result is an unprecedented ability to derive optimal asset allocations quickly and accurately. For instance, QAOA can account for intricate dependencies between assets, market volatility, and real-time constraints, offering financial analysts a robust tool to make informed decisions. Furthermore, this capability extends beyond static analysis; quantum optimization seamlessly integrates real-time market data, enabling portfolios to adjust to sudden shifts or emerging opportunities in the financial landscape. This adaptability enhances efficiency and empowers firms to maintain a competitive edge in an ever-changing market environment.

RISK ANALYSIS

Understanding and managing risk is paramount in finance, as it directly influences decision-making and stability within the sector. Traditional risk models, such as Value at Risk (VaR) or Conditional Value at Risk (CVaR), rely heavily on historical data and statistical assumptions, often failing to capture the intricate and dynamic interdependencies of modern financial markets. This limitation becomes particularly evident during volatile periods, such as economic downturns or rapid market fluctuations, where traditional methods may overlook potential vulnerabilities.

Quantum computing offers a transformative approach to risk analysis by enabling the simulation of complex market scenarios that are computationally prohibitive for classical systems. For instance, quantum Monte Carlo simulations leverage the unparalleled processing power of quantum processors to model uncertainties with unprecedented precision. These simulations enable institutions to analyze a broader range of market conditions, including rare yet high-impact events such as financial crises, with a level of accuracy previously unattainable.

Beyond crisis prediction, quantum-enhanced simulations provide deeper insights into portfolio diversification strategies. Quantum systems enable financial firms to identify optimal strategies for mitigating systemic risks and enhancing portfolio resilience by simultaneously modeling the interactions between various assets and market variables. This advanced capability ensures institutions respond effectively to potential threats and proactively design strategies to thrive in uncertain markets.

FRAUD DETECTION

Financial fraud remains a significant challenge, costing businesses and consumers billions annually. Detecting fraudulent transactions involves identifying subtle, often imperceptible patterns within vast datasets, posing a considerable computational challenge. Classical systems approach this problem by analyzing one data point at a time, a method that can be slow and prone to missing complex correlations. Quantum computing fundamentally transforms this process by leveraging its ability to analyze vast amounts of data simultaneously through superposition and entanglement. This capability allows quantum systems to uncover hidden anomalies and patterns with unparalleled efficiency and speed. For instance, a quantum computer can identify intricate transactional relationships and behavioral irregularities across millions of data points in seconds, dramatically reducing the time required to detect fraudulent activities. As a result, institutions equipped with quantum-powered fraud detection systems can respond to threats in real-time, ensuring a higher degree of security and trust in financial operations.

Quantum machine learning (QML) algorithms, such as quantum support vector machines and quantum neural networks, can be applied to fraud detection. These algorithms can identify fraudulent patterns in real-time by analyzing transaction histories, customer behaviors, and network anomalies. The result is a significant reduction in false positives and a faster response to suspicious activities. Financial institutions leveraging quantum-powered fraud detection systems can offer enhanced security to their clients while reducing operational losses.

IMPACT ON THE FINANCIAL SECTOR

Integrating quantum computing into finance marks a monumental leap forward, fundamentally altering traditional practices and introducing groundbreaking levels of efficiency and innovation. Quantum technologies address some of the most formidable challenges that financial institutions face, including handling vast datasets, optimizing complex operations, and enhancing predictive capabilities. These technologies streamline processes requiring extensive computational power and time by dramatically reducing operational costs. Moreover, quantum computing enables superior decision-making by providing deeper insights into market trends, customer behaviors, and financial risks. This allows institutions to develop cutting-edge financial products tailored to specific needs, opening up avenues for services that were once thought unattainable. These innovations promise not only to transform internal operations but also to redefine the customer experience, enhancing the overall competitiveness of the financial sector.

- Algorithmic Trading: Quantum algorithms redefine the landscape of algorithmic trading by processing vast amounts of market data at unparalleled speeds. These algorithms can analyze intricate market trends, detect patterns, and execute trades precisely, providing firms with a competitive edge in high-frequency trading environments.

The ability to react in real-time to market shifts ensures institutions remain ahead of competitors, optimizing profits and minimizing risks.

- Risk Management: Quantum-enhanced modeling tools revolutionize risk management practices by providing unparalleled accuracy in capital allocation and regulatory compliance. Financial institutions can simulate complex market dynamics, stress-test portfolios, and accurately predict the impact of rare economic events. This advanced capability empowers firms to mitigate risks more effectively and ensure robust financial stability.
- Customer Insights: Quantum data analysis's capabilities extend beyond operational efficiencies to enhance customer experiences. Quantum systems can uncover nuanced insights into customer behaviors, preferences, and financial patterns by processing and analyzing vast datasets. These insights enable institutions to design highly personalized financial products and services, fostering deeper customer engagement and loyalty.

The cumulative impact of these advancements underscores the potential of quantum computing to reshape the financial sector, driving innovation and redefining the competitive landscape.

Moreover, the competitive landscape of finance is likely to undergo significant shifts. Early adopters of quantum computing will gain a strategic advantage, attracting more clients and investment opportunities. Conversely, firms that delay adoption risk falling behind, especially as quantum technology becomes more accessible.

CHALLENGES AND ETHICAL CONSIDERATIONS

Despite its promise, integrating quantum computing into finance presents several challenges. Developing quantum algorithms and hardware is expensive, and the technology is still nascent. Additionally, the shift to quantum-powered systems raises ethical concerns, such as the potential for unequal access and exacerbating existing inequalities within the financial system.

Another critical consideration is cybersecurity. While quantum computing offers tools for enhanced security, it also poses a threat to traditional encryption methods. Financial institutions must invest in quantum-safe cryptographic solutions to safeguard sensitive data against potential quantum-enabled cyberattacks.

Quantum computing represents a paradigm shift for the financial sector, ushering in a new era of computational efficiency and innovation. By addressing some of the most intricate challenges, such as optimizing large-scale portfolios, assessing complex risk landscapes, and detecting sophisticated fraud patterns, quantum technologies promise to transform the way financial institutions operate. These advancements are not merely incremental but transformative, enabling previously unattainable solutions with classical computing systems. While the journey to full-scale quantum integration is fraught with technical and ethical challenges, the potential rewards far outweigh the risks. Quantum computing can enhance decision-making, reduce operational inefficiencies, and create novel financial products that cater to increasingly dynamic global markets. As this groundbreaking technology continues to evolve, early adopters will likely set themselves apart, leading the charge in redefining the competitive financial landscape. By embracing quantum innovation, they pave the way for a more resilient, secure, and forward-thinking industry, one that is poised to meet future demands.

CHAPTER 8

LOGISTICS, SUPPLY CHAINS, AND OPTIMIZATION

The logistics and supply chain sector is the backbone of global trade and commerce, orchestrating the seamless movement of goods, resources, and information across interconnected networks. This intricate and vital system ensures that products are delivered efficiently to consumers worldwide, underpinning the functioning of modern economies. However, it faces relentless challenges that require constant innovation and adaptability.

Influenced by seasonal trends, evolving market dynamics, and global events, fluctuating consumer demands place immense pressure on supply chains to remain agile. Additionally, the resilience of these networks is regularly tested by disruptions such as natural disasters, geopolitical tensions, and unforeseen crises like pandemics, which can cause significant delays and inefficiencies. Managing these complexities is further complicated by the global scale of logistics, where interdependent transport routes, warehouse networks, and cross-border regulations converge. This complexity necessitates advanced strategies to strike a balance between cost-effectiveness, reliability, and speed.

Compounding these operational challenges is the growing emphasis on sustainability. Companies are increasingly compelled to adopt greener practices to reduce their carbon footprints while maintaining or enhancing service levels. This includes optimizing transportation to cut emissions, minimizing waste in inventory management, and transitioning to renewable energy sources for powering operations.

Amidst these pressing challenges, quantum computing emerges as a revolutionary technology that potentially transforms the logistics and supply chain landscape. Unlike classical computing systems, which process information sequentially, quantum computing leverages the principles of quantum mechanics to process vast and complex datasets simultaneously. This enables the rapid evaluation of multiple variables and scenarios in real-time, making it an ideal tool for addressing the multifaceted problems faced by modern supply chains.

By enabling logistics companies to develop more innovative, more efficient routing, inventory management, and demand forecasting strategies, quantum computing promises to drive innovation and foster resilience and sustainability across global supply chains. Its transformative capabilities herald a new era of operational excellence, setting the stage for the next leap in logistics efficiency and reliability.

QUANTUM SOLUTIONS FOR ROUTING AND INVENTORY

The logistics and supply chain industry, which serves as the backbone of global commerce, grapples with numerous challenges related to efficiency, reliability, and scalability. Coordinating vast networks of transportation routes, warehouses, and inventory levels requires solving intricate optimization problems that often exceed the computational limits of classical systems. Fluctuating consumer demand, unexpected disruptions, and environmental considerations further exacerbate these challenges. Quantum computing, with its unparalleled ability to process complex calculations at extraordinary speeds, presents a transformative solution. By leveraging quantum mechanics, this technology can simultaneously evaluate vast combinations of variables, providing logistics companies with faster, more precise strategies to enhance operations and meet the ever-growing demands of global trade.

At the heart of logistics and supply chain optimization lies the need to address complex challenges, such as the Traveling Salesman Problem (TSP) and the Vehicle Routing Problem (VRP). The TSP involves determining the shortest route that visits a set of locations once and returns to the starting point. At the same time, the VRP extends this by incorporating factors like multiple vehicles, delivery time windows, and varying capacities.

These problems are crucial for the efficient operation of delivery systems. Still, they become increasingly difficult to solve as the number of locations and constraints grows, leading to an exponential rise in computational demands. Classical algorithms often fall short under such conditions due to their linear approach to processing potential solutions. In contrast, quantum computers utilize principles such as quantum superposition, which allows them to represent and process numerous possibilities simultaneously, and entanglement, which enables efficient interconnections between qubits. These quantum properties allow the evaluation of a vast array of potential solutions simultaneously, drastically reducing the time required to identify the optimal routing or scheduling strategies.

One promising approach is quantum annealing, which is particularly suited for optimization tasks. Companies like D-Wave Systems have successfully applied quantum annealing to routing problems early. For example, a quantum algorithm could optimize delivery routes for a fleet of trucks, taking into account variables such as traffic, delivery windows, and fuel efficiency. This task could take hours or even days for classical computers to solve. Inventory management, another critical aspect of supply chains, also benefits from quantum solutions. Maintaining optimal stock levels requires accurate demand forecasting and real-time adjustments to minimize costs and prevent shortages or overstocking. Quantum algorithms can analyze vast datasets and identify patterns that inform these decisions. By incorporating variables such as seasonal trends, consumer behavior, and supplier lead times, quantum-powered systems enable companies to manage their inventory more dynamically and efficiently.

REAL-WORLD EXAMPLES IN TRANSPORTATION AND DELIVERY

The transformative potential of quantum computing is already being explored in real-world logistics applications, where its ability to tackle complex optimization challenges is being tested. Several industry leaders are partnering with quantum computing firms to address longstanding issues like inefficient routing, high operational costs, and demand forecasting. These collaborations aim to leverage quantum algorithms to redefine how goods are transported, stored, and delivered, offering a competitive edge and paving the way for more sustainable and efficient logistics networks.

1. DHL and Quantum Optimization: The global logistics giant DHL has been at the forefront of experimenting with quantum computing to optimize its supply chain operations. By collaborating with quantum technology providers, DHL has tested quantum algorithms to enhance warehouse operations by identifying the most efficient storage layouts and retrieval processes. Additionally, quantum systems have refined vehicle routing, taking into account complex variables such as real-time

traffic data, delivery time windows, and vehicle fuel efficiency. Early results suggest that these innovations could reduce overall transportation costs, significantly decrease fuel consumption, and improve delivery times, resulting in both financial and environmental benefits.

2. Volkswagen's Traffic Flow Optimization: In a groundbreaking project, Volkswagen utilized quantum computing to optimize urban traffic flow. The company used a quantum algorithm to determine the most efficient routes for its fleet of vehicles, leveraging real-time data on road congestion and traffic patterns. This approach improved travel times and reduced fuel consumption by minimizing idle times and detours. While initially designed for urban traffic management, this technology shows great promise for supply chain logistics, where optimizing the routing of delivery vehicles is critical for timely and cost-effective operations.

3. FedEx's Quantum-Driven Routing Systems: FedEx, a global leader in package delivery, has made significant investments in exploring quantum computing to enhance its routing systems. With millions of packages to handle daily across a vast international network, FedEx faces a logistical challenge that traditional algorithms struggle to manage efficiently. By leveraging quantum algorithms, FedEx can quickly process vast datasets, enabling real-time route adjustments that take into account changing weather, traffic, and package priorities.

These quantum-driven systems have the potential to enhance customer satisfaction by ensuring faster and more reliable deliveries while also reducing operational costs.

4. Optimization in Smart Ports: Smart ports, like those in Singapore and Rotterdam, are pioneering the adoption of quantum-inspired technologies to revolutionize container shipping operations. These ports face immense challenges in scheduling the movement of ships, cranes, and containers to maximize throughput and minimize delays. Quantum computing has shown promise in solving these complex scheduling problems by considering multiple constraints simultaneously, such as ship arrival times, cargo priorities, and dock availability. The result is a more efficient port operation, which reduces turnaround times and enhances overall productivity. Such advancements position quantum computing as a crucial tool for modernizing global trade hubs.

FUTURE OUTLOOK

The integration of quantum computing into logistics and supply chains is still in its early stages, but the potential benefits are immense. As quantum hardware matures and becomes more accessible, industries are poised for a transformative shift. Quantum solutions promise to enhance last-mile delivery efficiency by dynamically optimizing routes under fluctuating conditions, such as traffic and weather. For cross-border logistics, quantum algorithms could streamline customs processes by analyzing complex trade regulations and predicting clearance times more accurately. Supply chain resilience could also be bolstered by leveraging quantum-powered predictive models to anticipate and mitigate disruptions such as demand surges or material shortages. These capabilities could unlock a new era of efficiency and reliability across global supply chains.

Moreover, the environmental implications of quantum-enhanced logistics cannot be overlooked. By utilizing quantum algorithms to optimize delivery routes and streamline inventory management, companies can substantially reduce fuel consumption and waste resulting from overproduction or excess inventory. This not only lowers operational costs but also contributes to sustainability goals. Such advancements align with global efforts to mitigate climate change by decreasing greenhouse gas emissions and supporting ambitious carbon reduction targets, ensuring a more sustainable future for logistics and beyond.

CHALLENGES AND CONSIDERATIONS

While the promise of quantum computing in logistics is undeniable, several challenges remain. Quantum computers, although groundbreaking, are still in their early stages of development. Current hardware limitations, including noise—unwanted disturbances that affect quantum operations—and decoherence—the loss of quantum state stability over time—significantly restrict their practical applications in large-scale scenarios. Furthermore, integrating these cutting-edge quantum solutions into existing supply chain systems is not straightforward; it demands considerable financial investment, specialized expertise, and the adaptation of legacy systems to accommodate quantum innovations. These barriers underscore the need for a gradual adoption and ongoing research to realize the potential of quantum technologies in logistics fully.

Despite these hurdles, the rapid pace of innovation in quantum technology suggests that these barriers will diminish over time. Continuous advancements in hardware, such as reducing noise and improving error correction, will make quantum computers more reliable and scalable. Companies are already experimenting with pilot projects, applying quantum algorithms to optimize supply chains, improve routing, and enhance inventory forecasting. These early successes demonstrate the vast potential for quantum solutions to overcome logistical inefficiencies and redefine industry standards. As these technologies mature, the logistics sector is poised for a profound transformation, unlocking previously unattainable capabilities.

Quantum computing is poised to revolutionize logistics and supply chains by solving complex optimization problems quickly and accurately. The applications are vast and transformative, from route planning to inventory management. Real-world examples, such as those from DHL, Volkswagen, and FedEx, highlight the potential for quantum solutions to enhance efficiency, reduce costs, and support sustainability. As technology continues to evolve, it will undoubtedly become a cornerstone of modern logistics, shaping the future of global commerce.

CHAPTER 9

STRATEGIC OPPORTUNITIES FOR BUSINESSES

Strategic opportunities in quantum computing give businesses a rare chance to establish themselves as pioneers in a transformative technological era. This is not just about adopting a new computational tool but about reimagining how organizations can operate, innovate, and deliver value in unprecedented ways. To seize this opportunity, businesses must undertake deliberate and comprehensive steps to understand the unique capabilities of quantum computing and align these capabilities with their broader goals. At its core, quantum computing is designed to solve problems beyond the reach of classical systems, such as intricate optimization challenges, the development of unbreakable cryptographic protocols, and the simulation of complex molecular and physical interactions. For businesses, recognizing and capitalizing on where these quantum advantages intersect with their unique needs lays the foundation for creating a robust and forward-looking strategy.

Understanding the capabilities of quantum computing requires organizations to think beyond traditional boundaries. Unlike classical computing, which relies on binary bits to process information, quantum computing harnesses phenomena such as superposition and entanglement to process vast amounts of data simultaneously. This enables it to tackle computationally intensive problems that classical computers would struggle with or require decades to solve. For example, in industries such as logistics, quantum algorithms can significantly optimize supply chains, ensuring that inventory and routing decisions are efficient and adaptable to real-time changes. Similarly, in the pharmaceutical industry, quantum computing holds the potential to simulate molecular behaviors with unprecedented accuracy, thereby accelerating drug discovery and reducing the costs associated with trial-and-error research.

The alignment of quantum capabilities with organizational goals goes beyond identifying immediate applications. It involves envisioning long-term, transformational outcomes. Businesses must ask how quantum computing can enhance current operations and explore how it can unlock entirely new business models and revenue streams. For instance, a manufacturing firm might leverage quantum algorithms to design novel materials that enable the development of next-generation products. At the same time, a financial institution might utilize quantum-driven risk analysis to offer more sophisticated and customized investment solutions. By embedding quantum opportunities into their strategic vision, businesses can position themselves as industry leaders, driving innovation while creating sustainable competitive advantages.

Building awareness across leadership and operational teams is essential. Decision-makers need a clear grasp of quantum computing's strengths and limitations to ensure that any investment or integration is purpose-driven. For example, a logistics firm may explore how quantum algorithms can optimize supply chains, while a pharmaceutical company might delve into quantum simulations to accelerate drug discovery. These explorations must be guided by a focused vision of how quantum solutions can create tangible value.

Another pivotal step is cultivating specialized expertise within an organization. Quantum computing operates at the intersection of quantum mechanics, advanced mathematics, and computer science, requiring a highly skilled workforce. Businesses can achieve this by recruiting professionals well-versed in quantum technologies, partnering with academic institutions, and investing in training for their existing employees. By fostering a culture of continuous learning, organizations can position themselves as frontrunners in the quantum space.

Research and development (R&D) investment plays a central role in this transition. Businesses must allocate resources to experiment with quantum algorithms, understand how to integrate them into existing systems, and develop proprietary solutions tailored to their unique challenges. Early engagement in R&D enables companies to test the waters, identify practical use cases, and prepare for larger-scale adoption as the technology matures. These investments are not merely expenditures but are strategic moves toward long-term innovation and competitiveness.

Security is another critical area where quantum computing presents both opportunities and challenges. The ability of quantum computers to crack traditional encryption methods is a looming threat, necessitating immediate action to adopt quantum-resistant cryptographic protocols. Businesses that proactively secure their data and systems will protect themselves and gain trust from customers and stakeholders in an increasingly security-conscious market.

Engaging with the broader quantum ecosystem is critical to preparing for the quantum era. Governments, industry consortia, and regulatory bodies are at the forefront of shaping this landscape, influencing the development of global standards, regulations, and research priorities. By actively participating in these initiatives, businesses can stay ahead of emerging trends, contribute to the formulation of policies, and ensure their strategies align with the evolving technological and regulatory framework.

Participation in the quantum ecosystem enables businesses to access cutting-edge research and technology developments. This involvement can take various forms, such as collaborating with academic institutions, joining quantum-focused industry groups, or partnering with technology providers leading the charge in quantum advancements. These collaborations offer businesses early access to new tools and methodologies, fostering a network of shared expertise that drives collective progress.

Furthermore, businesses that take an active role in shaping quantum policies can influence the direction of the industry. For example, engaging with regulatory bodies allows organizations to advocate for standards that address their specific needs or concerns, such as ensuring the interoperability of quantum systems across different platforms. This proactive approach helps businesses avoid being caught off guard by policy changes while positioning themselves as thought leaders in the quantum space.

Additionally, aligning with the quantum ecosystem helps companies mitigate risks associated with rapid technological advancements. By staying closely connected to industry developments, businesses can anticipate disruptions and adapt their strategies accordingly. For instance, early engagement in discussions about quantum-safe cryptography ensures that organizations are prepared to secure their data against future quantum threats, safeguarding their operations and customer trust.

Engaging deeply with the quantum ecosystem also provides opportunities for innovation and collaboration that transcend individual company boundaries. Businesses can co-create solutions that address complex challenges, accelerate technological progress, and unlock new market opportunities by working alongside other stakeholders. This cooperative spirit benefits the participating organizations and contributes to the broader goal of advancing quantum technology for the benefit of society and the economy.

Ultimately, active involvement in the quantum ecosystem enables businesses to transform from passive observers to influential contributors in the quantum revolution. By forging partnerships, participating in policy development, and staying attuned to industry trends, companies can ensure they are not merely reactive to change but are actively shaping the future of quantum technology and its application across the global marketplace.

Businesses must approach the quantum era with a long-term perspective. Quantum computing is still in its early stages, and its full potential will unfold over time. Companies that demonstrate patience, resilience, and adaptability will be best positioned to capitalize on future advancements. This means setting realistic goals, learning from incremental progress, and continuously refining strategies as the technology evolves.

In sum, strategic opportunities in the quantum era extend beyond adopting a novel technology; they demand a transformative approach to how businesses operate, innovate, and remain competitive in an ever-evolving landscape. Quantum computing is not merely a tool for solving complex problems but a catalyst for reimagining operational models, redefining value propositions, and creating new markets. By adopting a clear and deliberate strategy, organizations can not only unlock efficiencies and solve problems once thought insurmountable but also establish themselves as trailblazers in their respective industries, setting benchmarks for innovation and excellence.

Embracing quantum computing entails a paradigm shift in business strategy. It requires companies to think expansively about how quantum capabilities can integrate with existing processes while opening avenues for disruptive advancements. For instance, organizations can use quantum computing to refine optimization algorithms, enhance predictive modeling, or explore revolutionary approaches to R&D. This integration fosters a more profound synergy between technology and business objectives, enabling companies to achieve unprecedented levels of efficiency and innovation.

Moreover, the competitive advantage offered by quantum computing extends beyond immediate operational gains. Companies that pioneer the adoption of quantum technology stand to influence industry standards, attract top-tier talent, and establish themselves as thought leaders in the quantum space. These early movers will not only lead in deploying quantum solutions but also shape the narrative around the technology's transformative potential, thereby reinforcing their market position and fostering long-term growth.

Strategic foresight and adaptability are critical for businesses that capitalize on quantum opportunities. Rapid advancements and an uncertain trajectory characterize the quantum era, necessitating a mindset of continuous learning and resilience. Organizations must be prepared to navigate unforeseen challenges, pivot as needed, and leverage their growing expertise to remain at the forefront of innovation. This proactive and adaptive approach will enable businesses to turn potential disruptions into opportunities for growth and differentiation. In embracing quantum computing, businesses are not just investing in technology but embarking on a journey of transformation that will shape the future of industries on a global scale. By strategically leveraging quantum capabilities, they can unlock unparalleled opportunities, redefine what is possible, and lead their fields into a new era of technological and operational excellence.

HOW BUSINESSES CAN PREPARE FOR THE QUANTUM ERA

The dawn of the quantum era is poised to redefine industries, presenting unprecedented challenges and remarkable business opportunities. Companies that anticipate and adapt to this transformation will stand to gain significant competitive advantages. Preparing for the quantum era requires strategic planning, investment in talent and technology, and fostering an organizational culture that values innovation and adaptability. One of the first steps for businesses is to build awareness and understanding of quantum computing. This means not only recognizing its potential but also understanding its current limitations. Quantum computing is not a universal replacement for classical computing; instead, it excels in solving specific types of problems, such as optimization, cryptography, and complex simulations. Leaders must identify areas within their operations where quantum computing could provide a significant edge. For example, a logistics company might explore quantum algorithms for supply chain optimization, while a pharmaceutical firm could leverage quantum simulations for drug discovery.

Another critical step is talent acquisition and development. Quantum computing requires specialized knowledge of quantum mechanics, computer science, and mathematics. Organizations can begin by hiring experts in these fields and investing in the upskilling of their existing workforce. Collaborations with universities and research institutions can help businesses stay at the forefront of quantum innovation. Additionally, companies should explore partnerships with technology providers offering quantum solutions, such as IBM, Google, and Microsoft, to gain access to quantum hardware and software platforms without needing substantial in-house infrastructure.

Investing in research and development (R&D) is crucial for businesses seeking to leverage the potential of quantum computing. This includes experimenting with quantum algorithms, developing proprietary applications, and understanding how to integrate quantum solutions with existing systems. Early investments in R&D can provide businesses with a first-mover advantage, enabling them to capture market share in emerging quantum-enabled sectors. Furthermore, fostering an innovation-driven culture is vital. Employees should be encouraged to think creatively and experiment with quantum technologies to uncover new applications and solutions.

Businesses must also prepare for the potential cybersecurity challenges that quantum computing poses. Quantum computers can break traditional encryption methods, such as RSA and ECC, which are widely used to secure sensitive data. To mitigate this risk, organizations should begin transitioning to quantum-resistant cryptographic protocols. This proactive approach will safeguard data and ensure compliance with emerging cybersecurity standards in the quantum era.

Governments and regulatory bodies will likely play a significant role in shaping the quantum landscape. Businesses should actively monitor policy developments and engage in industry groups to influence regulations and standards. By participating in the broader quantum ecosystem, companies can help shape a favorable environment for innovation while ensuring they remain compliant with new requirements.

Businesses must adopt a long-term perspective to navigate the quantum era effectively. Quantum computing, while promising, is still in its formative stages, and the path to widespread adoption will require both technological breakthroughs and the maturation of its ecosystem. Companies approaching this journey with patience and a forward-thinking mindset will be better equipped to thrive.

A key element of this long-term perspective is the ability to set realistic goals. Businesses must acknowledge that quantum computing is not an overnight solution but rather a gradual evolution of capabilities. This involves breaking down quantum initiatives into manageable phases, each focused on achieving incremental progress. For instance, early stages involve exploratory pilot projects or proof-of-concept experiments to assess the technology's applicability to specific business needs. Such efforts help organizations build foundational knowledge while minimizing risks associated with larger-scale implementations.

Continuous learning is another cornerstone of long-term success. As quantum computing advances, new algorithms, hardware innovations, and applications will emerge, necessitating that businesses remain informed and adaptable. This can be achieved through partnerships with academic institutions, industry consortia, and engagement with quantum technology providers. By fostering an organizational culture that prioritizes education and innovation, companies can remain agile and responsive to emerging opportunities.

Moreover, businesses must cultivate resilience in the face of uncertainty. The road to quantum maturity will inevitably involve setbacks, whether in technical limitations, regulatory hurdles, or slower-than-anticipated progress. Companies that view these challenges as opportunities for growth and refinement will be better positioned to sustain momentum. Resilience also means preparing for potential disruptions within their industries. Quantum computing's transformative power could reshape competitive landscapes, making it essential for businesses to reassess their strategies and adapt to new realities continually.

Ultimately, adopting a long-term perspective means embracing the potential of quantum computing while acknowledging its complexities. Companies that commit to incremental progress, prioritize continuous learning, and maintain resilience will be best positioned to capitalize on the quantum revolution. In doing so, they will drive innovation within their organizations and contribute to shaping the future of industries on a global scale.

CASE STUDIES OF EARLY ADOPTERS

To illustrate the strategic opportunities that quantum computing presents, let us examine case studies of early adopters who are already leveraging this transformative technology. These early adopters span the automotive, finance, aerospace, and material science industries. Each sector faces unique challenges that quantum computing effectively addresses, from optimizing complex logistics networks to accelerating research and development processes. This chapter highlights how these trailblazing organizations strategically apply quantum solutions to unlock unprecedented efficiencies and drive innovation.

VOLKSWAGEN: OPTIMIZING TRAFFIC FLOW

Volkswagen has emerged as a pioneer in applying quantum computing to optimize urban traffic flow. In collaboration with D-Wave, the company developed a quantum algorithm capable of analyzing vast real-time traffic data. This solution aims to reduce congestion, minimize travel times, and enhance overall mobility in smart cities. By investing in quantum technology early, Volkswagen has positioned itself as a leader in the automotive and mobility sectors, demonstrating the practical benefits of quantum-enabled solutions.

JP MORGAN CHASE: FINANCIAL MODELING

JP Morgan Chase is exploring quantum computing in the financial sector to improve risk analysis and portfolio optimization. Using IBM's quantum platform, the company has developed quantum algorithms for solving complex optimization problems that are beyond the reach of classical computing. These advancements can revolutionize financial modeling, enabling more accurate predictions and better decision-making. By integrating quantum computing into its operations, JP Morgan Chase sets a new standard for innovation in the financial industry.

BASF: ACCELERATING CHEMICAL RESEARCH

Chemical giant BASF is leveraging quantum computing to accelerate research and development in material science. The company aims to design more efficient catalysts and develop new materials for energy storage and other applications by simulating molecular interactions at a quantum level. BASF's partnership with quantum computing startups underscores the value of collaboration in driving innovation. Their early adoption of quantum technology accelerates their R&D efforts and positions them as leaders in sustainable solutions.

AIRBUS: OPTIMIZING AIRCRAFT DESIGN

Airbus uses quantum computing to optimize aircraft design and performance in the aerospace industry. Quantum algorithms allow engineers to simulate aerodynamic properties with unprecedented precision, reducing the time and cost associated with prototyping. Airbus's quantum initiatives highlight the potential for quantum computing to revolutionize complex engineering processes, ensuring safer, more efficient, and environmentally friendly aviation solutions.

ZAPATA COMPUTING: PARTNERING ACROSS INDUSTRIES

Zapata Computing, a quantum software company, has partnered with various industries to develop tailored quantum solutions.

Their Orquestra platform enables businesses to experiment with quantum workflows, integrating quantum and classical computing to tackle real-world challenges. By facilitating access to quantum technology, Zapata Computing empowers companies across sectors to explore the transformative potential of quantum computing without requiring extensive in-house expertise.

LESSONS FROM EARLY ADOPTERS

The experiences of these early adopters offer valuable lessons for businesses preparing for the quantum era. Collaboration stands out as one of the most critical strategies. By partnering with technology providers, startups, and academic institutions, businesses can accelerate their innovation cycles and access cutting-edge quantum technologies without bearing the entire cost or complexity of development alone. These partnerships often lead to the sharing of expertise and resources, lowering barriers to entry and fostering a robust innovation ecosystem.

A focused approach is equally essential. Instead of attempting to apply quantum computing broadly across all operations, successful companies identify specific use cases where quantum computing can deliver measurable benefits. For example, logistics organizations may utilize quantum algorithms to optimize supply chain networks, while financial institutions can apply quantum techniques to enhance portfolio optimization and risk management. This laser-focused application ensures that quantum efforts yield tangible results, making it easier to justify investments and build organizational momentum around the technology.

Adaptability and resilience are two key pillars for businesses navigating the rapidly evolving landscape. Quantum computing is a rapidly changing field, with breakthroughs and challenges emerging frequently. Companies must remain agile, adjusting their strategies as the technology matures. This could involve scaling up quantum initiatives as hardware becomes more reliable or exploring hybrid solutions that combine classical and quantum computing to address practical challenges. Resilience also means being prepared for setbacks, such as technical limitations or slower-than-expected development timelines, and viewing these as opportunities for learning and growth.

Moreover, businesses can draw inspiration from these pioneers in developing internal frameworks for integrating quantum computing. Early adopters often create interdisciplinary teams that combine domain experts with quantum specialists to bridge the gap between theory and application. They also establish pilot projects that allow them to test quantum solutions on a smaller scale before rolling them out across the organization. This incremental approach minimizes risks while maximizing the potential for success.

Ultimately, the lessons from early adopters underscore that preparing for the quantum era necessitates more than just technological investment. It demands a strategic mindset, a willingness to collaborate, and an ability to adapt to the unpredictable nature of this emerging field. By embracing these principles, businesses can position themselves not only to navigate but also to lead the quantum revolution.

The quantum era represents a once-in-a-generation opportunity for businesses to redefine their operations and create value in ways previously unimaginable. By embracing quantum computing, companies can unlock new efficiencies, innovate at unprecedented scales, and gain a competitive edge in the global marketplace. However, success requires a strategic and forward-thinking approach. Businesses that invest in talent, technology, and partnerships today will be tomorrow's leaders, shaping the future of the quantum revolution.

PART 4

QUANTUM COMPUTING AND EVERYDAY LIFE

CHAPTER 10

HEALTHCARE AND PERSONALIZED MEDICINE

THE QUANTUM REVOLUTION IN HEALTHCARE

Quantum computing is poised to revolutionize the healthcare industry profoundly, spanning genomics, drug discovery, precision medicine, and beyond. At the heart of these advancements lies the unparalleled potential of quantum systems to process and analyze vast amounts of complex biological data at unprecedented speeds. Traditional computational methods often fall short when addressing the scale, intricacy, and multidimensionality of tasks like decoding the human genome, modeling protein dynamics, or identifying optimal drug formulations. Quantum computing introduces a transformative approach, shattering existing boundaries and promising breakthroughs that could redefine how we diagnose, treat, and prevent diseases on a global scale.

This revolutionary technology leverages the principles of quantum mechanics, such as superposition and entanglement, to address the healthcare sector's most pressing challenges. It offers solutions previously deemed unattainable, from improving diagnostic accuracy to accelerating treatment development. By enabling simultaneous analysis of multiple possibilities, quantum computing transcends the linear limitations of classical systems, empowering medical professionals and researchers to explore opportunities with extraordinary speed and precision.

One of the most compelling aspects of quantum computing in healthcare is its ability to address problems previously considered intractable. For instance, quantum systems excel at managing complex combinatorial systems, enabling a holistic analysis of interconnected biological systems. This ability allows scientists to uncover relationships and insights that were previously obscured, paving the way for novel diagnostic tools and therapeutic solutions. Furthermore, the quantum paradigm facilitates real-time adaptability, enabling researchers to dynamically adjust models and experiments in response to emerging data, significantly accelerating the pace of discovery.

This paradigm shift is more than a technological advancement; it represents a fundamental change in how we approach healthcare challenges. Quantum computing has the potential to save countless lives by providing faster, more accurate diagnostics and treatments while simultaneously reducing costs and addressing global health challenges on an unprecedented scale. For instance, by optimizing resource allocation, quantum algorithms could ensure equitable distribution of medical supplies and services, particularly in underserved regions. Moreover, integrating quantum systems into healthcare could catalyze interdisciplinary collaborations, uniting medical, physics, data science, and ethics experts to create comprehensive solutions for humanity's most pressing health concerns. As the quantum revolution continues to unfold, its ripple effects are set to transform the healthcare landscape, offering a future where medical science and technology converge to create a healthier, more equitable world.

ROLE OF QUANTUM COMPUTING IN GENOMICS

Genomics, the study of an organism's complete set of DNA, has already revolutionized our understanding of diseases, genetic predispositions, and inherited traits. However, analyzing genomic data requires immense computational power due to its vast complexity. A human genome comprises approximately three billion base pairs, and decoding this massive dataset to uncover meaningful patterns represents a Herculean task for classical computers. With their unique ability to operate in superposition, Quantum computers offer a revolutionary approach by analyzing multiple possibilities simultaneously, making them particularly suited for the intricate demands of genomics.

The potential of quantum computing in genomics extends beyond mere data processing. Quantum systems enable researchers to identify genetic markers linked to diseases quickly and accurately. For instance, applying quantum algorithms can rapidly detect variations in genetic sequences that predispose individuals to conditions such as cancer, cardiovascular diseases, or neurological disorders. This capability facilitates earlier detection, enabling timely interventions that can significantly improve patient outcomes. Moreover, quantum computing empowers researchers to explore the genetic underpinnings of rare diseases more efficiently, providing a clearer understanding of previously difficult conditions to study.

This groundbreaking capability also transforms the field of personalized medicine, where treatments are customized to an individual's unique genetic blueprint. Quantum computing accelerates the identification of gene variations associated with specific conditions, guiding the development of more effective therapies that are tailored to minimize adverse effects. This individualized approach can revolutionize patient care, enabling clinicians to predict treatment responses more accurately and design interventions that closely align with a patient's genetic makeup.

Furthermore, quantum computing significantly reduces the time and cost associated with genomic analysis, making personalized medicine more accessible and practical for a broader population. By facilitating extensive and large-scale genomic studies, quantum systems empower researchers to decode complex population-level health trends, uncovering insights that drive groundbreaking advancements in public health strategies. These studies enable the identification of genetic markers on a scale that has never been possible before, guiding policies and initiatives aimed at disease prevention and healthcare equity.

Additionally, the predictive capabilities of quantum systems are revolutionizing preventive care. For example, by analyzing a patient's genetic predispositions to various conditions, healthcare providers can implement precise and early lifestyle or medical interventions to mitigate potential health risks before they escalate into serious illnesses. This proactive approach is bolstered by the ability of quantum systems to integrate and analyze real-time data from wearable devices, medical records, and environmental factors, offering a more holistic understanding of a patient's health.

Quantum computing can also streamline the complex process of identifying gene variations linked to specific conditions. This accelerates the development of targeted therapies, enabling medical professionals to treat diseases with unprecedented precision and effectiveness. Quantum systems democratize access to personalized medicine by significantly reducing costs and timelines, making it a viable option even for resource-constrained settings. Furthermore, this transformative technology is advancing predictive models for rare genetic conditions, creating new pathways for preventative care and early intervention that were previously unattainable.

As these capabilities mature, quantum systems are redefining how healthcare transitions from a reactive to a proactive paradigm. They are developing predictive tools that anticipate genetic risks and propose tailored prevention strategies for individuals and entire populations. This shift holds the potential to transform healthcare delivery, ensuring that interventions are timely, effective, and personalized. Integrating quantum computing into genomic research and preventative care represents a monumental leap forward, setting the stage for a future where health outcomes are vastly improved for all.

ADVANCING DRUG DISCOVERY AND DEVELOPMENT

Drug discovery is another area where quantum computing is expected to make a monumental impact, offering an entirely new paradigm for pharmaceutical research. The traditional process of developing new drugs is notoriously time-consuming and expensive, often taking over a decade and billions of dollars to bring a single drug to market. This lengthy timeline is mainly due to the labor-intensive and trial-and-error nature of identifying promising compounds, testing their efficacy, and optimizing formulations. Quantum computers, however, have the potential to revolutionize this process by simulating molecular interactions with remarkable accuracy, eliminating much of the inefficiency and uncertainty involved in drug discovery.

For instance, quantum systems can model complex molecular

structures and their interactions with proteins in the human body, allowing for the prediction of a drug's effectiveness and potential side effects long before clinical trials begin. By doing so, quantum computing can significantly reduce the preclinical phase of drug development, enabling researchers to concentrate their efforts on the most promising candidates. This capability accelerates the discovery process and increases the likelihood of success, thereby reducing costly failures that occur during later stages of development.

Moreover, quantum computers enable the exploration of vast chemical spaces that are virtually impossible to analyze using classical methods. Researchers can simultaneously simulate and evaluate millions of molecular combinations, uncovering novel compounds that might remain undiscovered. Companies like IBM and Google are already exploring quantum applications in pharmaceutical research, collaborating with industry leaders to tackle complex medical challenges. In one notable instance, a quantum algorithm was used to model the behavior of a lithium hydride molecule, showcasing the potential for simulating even more intricate molecules in the future.

Beyond traditional drug development, quantum computing holds promise for addressing some of medicine's most persistent challenges. For example, the ability to simulate protein folding—a critical process for understanding diseases like Alzheimer's and Parkinson's—could lead to groundbreaking treatments and cures for these debilitating conditions. Quantum simulations can also facilitate the design of personalized drugs tailored to an individual's genetic makeup, ushering in a new era of precision medicine.

Furthermore, quantum-enhanced approaches to drug discovery can contribute to global health equity. By lowering the cost and time associated with developing new therapies, quantum computing can make life-saving drugs more accessible to populations in low-resource settings. This transformative potential highlights the importance of integrating quantum technology into the pharmaceutical industry to advance medical science and ensure its benefits are widely accessible worldwide.

For example, quantum systems can model how a drug molecule interacts with proteins in the human body, predicting its effectiveness and potential side effects. This capability accelerates the discovery process and increases the likelihood of success. Companies like IBM and Google are already exploring quantum applications in pharmaceutical research, collaborating with industry leaders to tackle complex medical challenges. In one notable instance, a quantum algorithm was used to model the behavior of a lithium hydride molecule, demonstrating the potential for simulating more complex molecules in the future. Furthermore, the ability to simulate protein folding—a key to understanding diseases like Alzheimer's and Parkinson's—could lead to groundbreaking treatments and cures for these debilitating conditions.

REVOLUTIONIZING DIAGNOSTICS

Diagnostics is another critical area where quantum computing promises transformative advancements. Current diagnostic methods often rely on imaging and laboratory tests, which can be time-intensive and sometimes yield inconclusive results. Quantum computers can analyze patient data from multiple sources—including genetic information, medical histories, and real-time health metrics—to identify patterns and correlations that might elude classical systems.

For instance, quantum-enhanced machine learning algorithms can sift through vast datasets to detect anomalies indicative of diseases such as cancer or cardiovascular conditions. These algorithms recognize subtle patterns, enabling earlier and more accurate diagnoses. Furthermore, quantum computing can facilitate the integration of wearable health devices, allowing for continuous monitoring and real-time analysis of patient health. This opens the door to proactive healthcare, where potential issues are identified and addressed before they escalate into severe conditions. Quantum-assisted diagnostic tools are also expected to play a crucial role in combating pandemics by rapidly identifying and isolating new viral strains, potentially saving millions of lives.

PRECISION MEDICINE AND TAILORED TREATMENTS

Precision medicine, which focuses on customizing healthcare to individual patients, is another frontier for quantum computing. By leveraging quantum systems, researchers can develop treatment plans tailored to a patient's unique genetic, environmental, and lifestyle factors. For example, cancer treatments often involve a combination of surgery, chemotherapy, and radiation. Determining the optimal approach for each patient requires analyzing numerous variables, including tumor genetics and overall health.

Quantum computers can rapidly evaluate these variables, identifying the most effective treatment protocols with greater accuracy than classical systems. This capability also extends to chronic disease management. For conditions such as diabetes or autoimmune disorders, quantum computing can aid in designing personalized management plans that optimize medication, diet, and lifestyle interventions. The result is a more holistic approach to healthcare, improving patient outcomes and quality of life. Beyond individual cases, this technology is enabling a shift toward population-level healthcare optimization, where entire demographics benefit from tailored health strategies informed by big data.

ACCELERATING RESEARCH AND INNOVATION

Beyond individual applications, quantum computing is set to accelerate medical research. By enabling faster and more accurate data analysis, quantum systems can uncover previously out-of-reach insights. This includes identifying new disease biomarkers, understanding complex biological systems, and exploring novel therapeutic approaches. For example, researchers can utilize quantum simulations to investigate protein folding, a process crucial to understanding diseases such as Alzheimer's and Parkinson's.

Quantum computing also has the potential to revolutionize clinical trials, which are a cornerstone of medical research. Traditionally, clinical trials involve lengthy patient recruitment processes, data collection, and analysis. Quantum algorithms can optimize trial design by identifying the most promising candidates, predicting outcomes, and reducing the time required to gather meaningful results. This accelerates the development of new treatments, bringing life-saving innovations to patients more quickly. Furthermore, the ability to simulate and predict outcomes before trials begin could significantly reduce the cost and time investment needed, making groundbreaking therapies more accessible worldwide.

ADDRESSING GLOBAL HEALTH CHALLENGES

The implications of quantum computing in healthcare extend beyond individual patients to address global health challenges.

For instance, infectious diseases like COVID-19 have highlighted the need for rapid vaccine development and distribution. Quantum systems can model viral mutations and predict the efficacy of potential vaccines, enabling a faster response to outbreaks. Additionally, quantum computing can aid in optimizing supply chains for medical supplies, ensuring timely delivery to areas in need.

In resource-limited settings, quantum computing can support efforts to combat diseases like malaria and tuberculosis by identifying cost-effective treatment strategies. Furthermore, quantum-enhanced diagnostics can improve access to quality healthcare in underserved regions, bridging gaps in medical infrastructure. By addressing these challenges, quantum computing has the potential to significantly improve global health outcomes. Large-scale public health initiatives could also leverage this technology to model and predict the spread of diseases, enhancing prevention and response strategies.

ETHICAL CONSIDERATIONS AND CHALLENGES

While the potential benefits of quantum computing in healthcare are immense, it is essential to address ethical considerations and challenges comprehensively. Data privacy remains one of the most pressing concerns, as quantum systems can process and analyze vast amounts of sensitive patient information. This raises the stakes for cybersecurity, necessitating the implementation of robust encryption protocols that can withstand classical and quantum threats. Ensuring compliance with stringent regulations and building public trust through transparency and accountability will be crucial in mitigating concerns about data misuse.

Furthermore, the high cost and technical complexity of quantum technology present significant barriers to accessibility. These challenges exacerbate disparities in global healthcare systems, where resource-rich nations can leap ahead, leaving developing regions at a disadvantage. Addressing this inequity requires coordinated efforts to democratize access to quantum tools and knowledge, including initiatives to subsidize infrastructure and foster international collaboration. Governments, nonprofits, and private organizations must work together to ensure that quantum advancements benefit humanity universally, not just select populations.

Another significant hurdle lies in the seamless integration of quantum systems with existing healthcare infrastructure. This involves upgrading current systems to accommodate quantum capabilities and training healthcare professionals to utilize these advanced tools effectively. Developing intuitive interfaces and comprehensive training programs will be pivotal in bridging the gap between cutting-edge quantum technologies and practical healthcare applications. Moreover, addressing the technological divide between advanced and developing regions will require tailored strategies that consider the unique challenges and limitations of different healthcare environments.

Ethical frameworks must evolve in tandem with technological advancements to prevent and address potential misuse of quantum healthcare technologies. For instance, ensuring that quantum applications prioritize patient welfare over profit motives will be critical in maintaining ethical integrity. Collaborative efforts among governments, industry leaders, academic institutions, and ethics boards are crucial to establishing guidelines that safeguard against exploitation while promoting innovation.

Public engagement and education are vital in shaping a future where quantum computing is embraced as a transformative force for good. Raising awareness about its benefits, addressing misconceptions, and involving diverse stakeholders in decision-making processes will help foster an environment of trust and cooperation. By proactively addressing these challenges, the healthcare sector can unlock the full potential of quantum computing, ensuring that its benefits are distributed equitably and implemented ethically.

A QUANTUM-POWERED FUTURE FOR HEALTHCARE

The integration of quantum computing into healthcare marks a profound paradigm shift with the potential to save countless lives, dramatically reduce costs, and revolutionize patient experiences on a global scale. Quantum computing's promise lies in its unparalleled ability to tackle intricate challenges that classical systems struggle to address. From enhancing diagnostic accuracy and accelerating drug discovery to advancing personalized medicine and addressing global health disparities, quantum computing unlocks new possibilities for medical breakthroughs that were once considered impossible. These transformative solutions promise a future where innovations occur at unprecedented speeds and with extraordinary precision, reshaping the healthcare landscape in unimaginable ways.

At its core, quantum computing transcends the limitations of traditional computational methods by leveraging principles such as superposition and entanglement. These capabilities enable quantum systems to process and analyze vast amounts of data simultaneously, making them uniquely suited for addressing the multifaceted complexities of healthcare. For example, quantum algorithms can unravel patterns in genetic data, identify biomarkers for diseases with unparalleled accuracy, and simulate molecular interactions to develop life-saving drugs. These advancements enhance the efficiency of medical research and expand the horizons of what is achievable in modern medicine.

Beyond individual applications, the ripple effects of quantum computing are set to redefine healthcare systems globally. By optimizing resource allocation, quantum technologies can ensure equitable access to medical supplies and services, particularly in underserved regions. This technology also holds the potential to revolutionize large-scale public health initiatives, enabling the modeling and prediction of disease outbreaks with unprecedented accuracy. These capabilities could transform how governments and organizations respond to pandemics, offering proactive and practical solutions to mitigate the impact of global health crises.

The paradigm shift introduced by quantum computing is not just about technological advancements but about fostering interdisciplinary collaborations that unite experts from medicine, physics, data science, and ethics. These collaborations are crucial for creating holistic solutions to humanity's most pressing health concerns. By facilitating real-time adaptability, quantum systems empower researchers to dynamically refine models and experiments in response to emerging data, accelerating the pace of discovery and innovation. This adaptability ensures that healthcare evolves with the rapid advancements in science and technology.

As the quantum revolution unfolds, its implications for healthcare are profound and far-reaching. This transformative technology offers a future vision where medical science and cutting-edge technology converge to create a healthier, more equitable world. Integrating quantum systems into healthcare is not merely an enhancement but a fundamental reimagining of how we understand and address human health, setting the stage for breakthroughs that will resonate for future generations.

As we navigate this quantum-powered future, addressing ethical considerations becomes paramount. The transformative power of quantum technology must be harnessed responsibly to ensure its advancements benefit humanity. Robust data privacy frameworks and equitable access to this technology are crucial for preventing disparities and ensuring global adoption. Moreover, interdisciplinary collaboration among researchers, policymakers, and healthcare providers will be essential to unlocking its full potential and mitigating challenges that arise during integration. By fully embracing quantum technology, the healthcare industry can unlock unparalleled possibilities. From decoding the complexities of diseases like Alzheimer's to optimizing large-scale public health initiatives, the ripple effects of quantum advancements will extend far beyond individual patients. They have the potential to redefine entire systems of care, improve efficiency, and address pressing challenges in underserved regions. As quantum systems evolve, their role in modern medicine will become increasingly indispensable, reshaping our understanding of health and disease in profound and inspiring ways.

The journey into quantum-powered healthcare is just beginning. As quantum computing matures, it will pave the way for a healthier, more equitable world where the boundaries of medical science are pushed further than ever before. The quantum revolution promises a brighter, healthier future for patients, providers, and communities worldwide—a future that combines cutting-edge technology with the universal goal of improving human well-being. As quantum systems continue to evolve, they stand to become an indispensable part of modern medicine, reshaping our understanding of health and disease in ways we are only beginning to imagine.

CHAPTER 11

QUANTUM-ENHANCED ENERGY SOLUTIONS

Humanity is racing toward sustainable energy systems in the face of climate change and dwindling fossil fuel reserves. This transition is not just a matter of technology but a critical step in securing a sustainable future for the planet. While traditional computational methods have driven significant advancements in renewable energy and grid optimization, they encounter limitations when managing the enormous complexity of modern energy systems. For instance, predicting and accommodating the fluctuating nature of renewable energy sources or managing the increasing interconnectivity of global energy grids often stretches classical computing methods to their limits. With its unparalleled ability to process vast datasets, model intricate interactions, and provide near-instantaneous solutions to complex problems, Quantum computing promises to overcome these barriers and unlock unprecedented potential in energy solutions.

Quantum computers can simulate the behavior of entire energy networks, taking into account variables such as weather conditions, energy consumption patterns, and storage capacities. These capabilities provide a new level of precision and adaptability, enabling more informed decisions that optimize efficiency and sustainability. This chapter delves into how quantum computing can revolutionize renewable energy systems by improving forecasting and infrastructure placement, enhancing grid optimization through advanced algorithms, and accelerating the development of next-generation materials for batteries and solar panels. Each application represents a crucial leap toward achieving a carbon-neutral world, demonstrating the power of quantum computing to address some of humanity's most pressing energy challenges.

THE COMPLEXITY OF RENEWABLE ENERGY SYSTEMS

Renewable energy sources, such as wind, solar, and hydropower, are inherently variable and distributed, making their effective management a highly complex task. Balancing the unpredictable and fluctuating nature of these resources requires precise forecasting, optimization, and real-time decision-making that accounts for numerous interrelated factors. While undeniably powerful, classical computers often fall short of addressing the full scope of these challenges due to their inability to model the intricate interactions between dynamic weather patterns, shifting energy demands, and the varying capacities of energy storage systems.

Quantum computing offers a transformative solution by enabling simulations and optimizations that were previously deemed infeasible. With their capability to analyze vast datasets and explore countless variables simultaneously, quantum computers can provide unprecedented insights into energy systems. For example, they can model the impact of regional weather patterns on renewable energy outputs with remarkable precision, allowing for better integration of solar and wind energy into the grid. By refining these forecasts, quantum computing reduces the dependency on fossil fuel backups, paving the way for a greener energy future.

Moreover, quantum algorithms can optimize the placement and operation of renewable energy infrastructure at an unprecedented scale. For instance, they can determine the optimal locations for wind farms and solar installations by analyzing geographic and meteorological data, while considering both economic and environmental factors. This precision ensures that renewable energy investments deliver maximum efficiency and cost-effectiveness. In addition, quantum systems can simulate the complex interplay between distributed energy resources, storage technologies, and consumer demand, resulting in a more resilient and adaptive energy network that is better equipped to meet future challenges.

For example, quantum computers can dramatically enhance the accuracy of weather forecasting models by analyzing vast and intricate datasets of meteorological information that would overwhelm classical systems. These advanced forecasts enable energy planners to accurately predict renewable energy outputs, taking into account complex and fluctuating variables such as wind patterns, cloud movements, and solar irradiance. This capability reduces reliance on fossil fuel backups and minimizes energy production and distribution inefficiencies. Furthermore, quantum algorithms can optimize the placement and operation of renewable energy infrastructure at scales and levels of detail previously unattainable. By simulating various geographic and environmental conditions, quantum systems can determine the ideal sites for wind farms, solar arrays, and hydropower stations, ensuring that energy investments yield maximum output while minimizing costs.

Additionally, these optimizations extend to operational strategies, such as determining the best times to activate or scale back specific installations to align with real-time energy demands and storage capacities. This comprehensive approach exemplifies the transformative potential of quantum computing in reshaping the renewable energy landscape.

QUANTUM ALGORITHMS FOR GRID OPTIMIZATION

Energy grids are the backbone of modern civilization, serving as a crucial source of power for homes, industries, and transportation systems. However, they face significant challenges in integrating renewable energy sources. Traditional grids, designed for centralized power generation, now struggle with the decentralized and intermittent nature of renewables, such as solar and wind energy. While these sources are environmentally friendly, they introduce variability and unpredictability that complicate energy distribution and demand management. Quantum computing emerges as a transformative solution capable of addressing these complexities, ensuring stability, efficiency, and resilience in modern energy grids.

Quantum algorithms, such as quantum annealing, excel at solving complex optimization problems that classical methods struggle to solve. One key application is optimizing energy flow within a grid to minimize losses and balance supply with demand across diverse and distributed sources. By rapidly evaluating numerous scenarios, quantum computers can identify the most efficient configurations for energy distribution, taking into account variables such as transmission capacities, fluctuating renewable energy inputs, and real-time consumption patterns. This capability enables more effective microgrid management, which combines local renewable generation with advanced storage systems and demand-side management techniques.

Moreover, quantum computing can revolutionize the scheduling and coordination of energy resources. For instance, it can optimize when and how battery storage systems are charged and discharged, considering factors like peak demand periods, grid stability requirements, and cost efficiencies. Additionally, it can enhance the integration of electric vehicles (EVs) into the grid by determining optimal charging schedules and locations, thereby preventing bottlenecks and ensuring smooth energy flow. This holistic approach improves grid reliability and lays the groundwork for a more interconnected and sustainable energy infrastructure. As renewable energy adoption continues to grow, the potential of quantum computing to solve grid-related challenges becomes increasingly vital, signaling a new era of innovation in energy systems.

Additionally, quantum computing can enhance the scheduling of energy resources. For instance, it can determine the optimal times to charge and discharge battery storage systems, reducing costs and improving grid reliability. These advancements are critical for achieving a sustainable energy future where renewable sources dominate.

ADVANCING BATTERY TECHNOLOGY WITH QUANTUM SIMULATIONS

Efficient energy storage is crucial for mitigating the intermittency of renewable energy sources and serving as a cornerstone for reliable and sustainable energy systems. While progressively improving, current battery technologies continue to face persistent challenges in terms of capacity, lifespan, efficiency, and cost-effectiveness. These hurdles limit their ability to support the growing demands of renewable energy integration. Quantum computing is key to unlocking the next generation of battery materials and technologies by enabling precise simulations of chemical interactions at the atomic level. In this area, classical computing methods fall short.

One emerging area is the development of solid-state batteries, which have the potential to revolutionize energy storage by offering significantly higher energy density, improved safety, and enhanced durability compared to traditional lithium-ion batteries. Quantum computers can simulate the intricate behavior of solid electrolytes with unmatched precision, identifying materials that exhibit optimal conductivity, stability, and compatibility with other battery components. This ability to model atomic and subatomic interactions accelerates the discovery and fine-tuning of novel materials, significantly reducing the time and cost of experimental trial-and-error approaches.

In addition to improving conventional batteries, quantum computing paves the way for breakthroughs in advanced energy storage technologies, such as hydrogen storage systems and fuel cells. Hydrogen, a clean energy carrier with vast potential, faces challenges related to storage efficiency and scalability. Quantum simulations can model the complex molecular interactions involved in hydrogen storage, enabling researchers to design lightweight, compact, and efficient solutions. Similarly, quantum computing can guide the development of innovative materials for fuel cells, enhancing their performance and reliability for broader transportation and grid storage applications.

Moreover, quantum-driven advancements extend beyond the design of materials. These technologies can also optimize the manufacturing processes for batteries and storage systems, identifying cost-effective techniques that minimize material waste and energy consumption. This comprehensive approach highlights the multifaceted role quantum computing will play in advancing energy storage, ensuring that next-generation systems are efficient, sustainable, economically viable, and widely accessible.

In addition to batteries, quantum computing can aid in designing efficient materials for hydrogen storage and fuel cells. Hydrogen is a clean energy carrier with significant potential, but current storage methods are bulky and inefficient. By modeling molecular interactions with unprecedented accuracy, quantum computers can guide the development of lightweight and compact hydrogen storage solutions.

ENHANCING SOLAR PANEL EFFICIENCY

Solar energy is one of the most abundant and clean energy sources available to humanity, yet its efficiency remains a limiting factor that restricts its broader adoption and impact. Quantum computing has the potential to drive groundbreaking advancements in photovoltaic technology by uncovering new materials, refining their properties, and optimizing their performance for maximum energy capture and conversion.

One of the most exciting focus areas is the development of perovskite solar cells, which have emerged as a promising alternative to traditional silicon-based cells. These cells have already shown potential for higher efficiency and lower production costs, but their widespread adoption has been hindered by issues related to stability and scalability. Quantum simulations can play a pivotal role in overcoming these challenges by identifying perovskite compositions with enhanced durability and performance under real-world conditions. Through detailed atomic-level modeling, quantum computers can explore an infinite array of molecular structures, pinpointing the best combination of efficiency, stability, and cost-effectiveness.

Beyond material discovery, quantum computing can also revolutionize the structural design of solar panels. Quantum systems excel at modeling the behavior of light-absorbing materials at the nanoscale, guiding the creation of innovative structures that maximize photon capture and improve energy conversion rates. For example, researchers can utilize quantum algorithms to design nanostructures that minimize energy losses by enhancing the interaction between photons and electrons within solar cells.

Another transformative application lies in optimizing the manufacturing processes for solar panels. By analyzing vast datasets on material properties, fabrication techniques, and production variables, quantum algorithms can identify cost-effective methods to produce high-performance solar cells. This approach reduces production costs and ensures consistent quality and efficiency across large-scale manufacturing operations.

Furthermore, quantum computing can contribute to advancements in hybrid solar technologies, which combine photovoltaic cells with other energy generation methods, such as solar thermal systems. By optimizing the interplay between these components, quantum algorithms can help create systems that harness solar energy more efficiently, paving the way for breakthroughs in energy generation and storage.

These advancements can potentially make solar energy more accessible and affordable globally. As quantum-enhanced solar technologies become more efficient and cost-effective, they can significantly accelerate the transition to renewable energy, thereby reducing fossil fuel dependency and mitigating the impacts of climate change.

Another application is optimizing the manufacturing processes for solar panels. Quantum algorithms can identify cost-effective methods to produce high-performance solar cells by analyzing vast datasets on material properties and fabrication techniques. These advancements can potentially make solar energy more accessible and affordable worldwide.

REAL-WORLD APPLICATIONS AND PROGRESS

Integrating quantum computing into the energy sector is already underway, marking a pivotal moment in the pursuit of sustainable and efficient energy systems. Companies and research institutions are exploring the transformative potential of this technology to address critical and long-standing challenges. For instance, IBM and D-Wave collaborate with energy providers to develop sophisticated quantum solutions to optimize grid operations. These collaborations demonstrate the practical application of quantum technologies in enhancing energy distribution networks and managing complex energy flows. Similarly, researchers are leveraging advanced quantum simulations to make groundbreaking progress in battery technologies and renewable energy materials, tackling limitations that have persisted for decades.

One notable example is the partnership between Volkswagen and quantum computing company Xanadu. This collaboration focuses on developing cutting-edge quantum algorithms designed to optimize the performance of electric vehicle batteries and energy management systems. This partnership is expected to accelerate advancements in electric mobility and sustainable transportation by addressing battery design and usage inefficiencies. Furthermore, such initiatives underscore the growing recognition of the far-reaching implications of quantum computing for the energy sector. From reducing energy losses to discovering new materials, the ongoing efforts by industry leaders signal a future where quantum computing is seamlessly integrated into energy innovations, driving both sustainability and efficiency at an unprecedented scale.

CHALLENGES AND THE ROAD AHEAD

Despite its promise, the application of quantum computing in energy solutions is still in its infancy. Challenges such as noise, decoherence, and the limited scalability of current quantum hardware must be addressed. Moreover, integrating quantum solutions into existing energy systems requires collaboration between technologists, policymakers, and industry stakeholders.

Continued investment in quantum research and development is essential to overcome these hurdles. Governments and private organizations must collaborate to develop the necessary infrastructure, train a skilled workforce, and promote innovation. Additionally, initiatives to standardize quantum computing tools and frameworks will accelerate their adoption in the energy sector.

A Quantum-Powered Energy Future

The potential of quantum computing to revolutionize energy systems is immense. From optimizing the integration of renewable energy to discovering breakthrough materials for batteries and solar panels, quantum technology offers solutions to some of the most pressing challenges in sustainable energy. By harnessing the power of quantum computing, humanity can accelerate the transition to a clean and resilient energy future.

As we continue to advance this field, ensuring that the benefits of quantum-enhanced energy solutions are distributed equitably is crucial. By prioritizing accessibility, collaboration, and sustainability, we can create a world where quantum computing powers our grids and empowers communities worldwide. This is the promise of a quantum-powered energy revolution—one that holds the potential to illuminate a brighter future for all.

CHAPTER 12

SMART CITIES AND THE INTERNET OF THINGS (IOT)

QUANTUM ADVANCES IN URBAN PLANNING AND IOT NETWORKS

In the age of rapid urbanization, cities worldwide face mounting challenges that threaten to undermine the quality of life and strain resources. From seemingly insurmountable traffic congestion and widespread energy inefficiencies to housing shortages and ineffective waste management systems, urban centers are grappling with increasingly complex problems. Traditional planning and management tools, reliant on classical computing, often prove inadequate for addressing these multifaceted issues at scale. The sheer volume of interconnected data produced by modern cities demands a paradigm shift in computational power and strategic decision-making. This is where the advent of quantum computing emerges as a revolutionary solution, offering capabilities far beyond those of classical systems. By harnessing the unique properties of quantum mechanics, quantum computing enables the processing of vast datasets, the simulation of intricate systems, and the optimization of urban operations in ways previously deemed impossible.

For example, imagine a city's traffic management system that responds to congestion in real-time and predicts it before it occurs, dynamically rerouting traffic to prevent delays altogether. Quantum computing's ability to process billions of variables simultaneously enables predictive and adaptive urban systems. Similarly, waste management could evolve into a precise and efficient system where collection schedules are optimized dynamically based on real-time sensor data, thereby reducing environmental impact and operational costs. These applications demonstrate how quantum technologies can fundamentally transform the planning and functionality of urban landscapes, paving the way for more intelligent, connected, and sustainable cities. By integrating quantum computing and IoT networks, cities have an unprecedented opportunity to enhance their infrastructure, promote sustainability, and improve the quality of life for their inhabitants.

THE ROLE OF QUANTUM COMPUTING IN URBAN PLANNING

Urban planning involves managing vast, interconnected systems, including transportation networks, energy grids, housing, and other related infrastructure. The complexity of these systems often makes it challenging for traditional tools to effectively address the myriad of demands and constraints posed by rapid urbanization. Quantum computing's ability to process vast datasets and simulate complex scenarios makes it an indispensable tool for modern urban planning.

For instance, city planners can leverage quantum algorithms to analyze real-time traffic patterns, identify persistent bottlenecks, and design more efficient, adaptive transportation systems. Imagine a scenario where a city's traffic management system identifies peak congestion points and reroutes vehicles preemptively, avoiding delays altogether. Unlike classical computers, which may take weeks or months to process such intricate computations, quantum systems can generate actionable insights within minutes, enabling swift and informed decision-making.

Beyond transportation, quantum computing offers groundbreaking applications in predictive modeling, a cornerstone of proactive urban management. Cities generate staggering amounts of data through IoT devices, sensors, and surveillance systems, creating a continuous stream of information on everything from environmental conditions to energy usage and population density. Quantum computing's unparalleled processing capabilities enable the rapid analysis of this data, allowing for the prediction of future trends with exceptional accuracy. These predictions can include urban growth patterns, housing demand, and shifts in energy consumption, enabling policymakers to design infrastructure that not only meets current needs but anticipates future ones, ensuring sustainable urban development.

Moreover, quantum computing can significantly enhance land-use planning. By integrating and analyzing a broad array of variables—including geographical data, environmental impacts, and economic constraints—quantum algorithms can identify optimal sites for new developments. This ensures that land-use decisions strike a balance between urban expansion and ecological preservation, thereby reducing urban sprawl and promoting more sustainable growth patterns. Additionally, these quantum-driven insights can inform zoning regulations and guide the expansion of public transportation, aligning urban growth with the principles of innovative and sustainable city initiatives.

Another critical aspect where quantum computing shines is its ability to simulate the effects of large-scale urban projects before implementation. These simulations can account for complex interactions between urban systems, such as how a new housing development might impact nearby traffic, water resources, and energy grids. By enabling planners to test multiple scenarios and outcomes, quantum computing minimizes the risks associated with large-scale urban projects, ensuring that resources are utilized efficiently and potential problems are mitigated well in advance.

In essence, quantum computing elevates urban planning to a previously unimaginable level of precision and foresight. It enables cities to adapt dynamically to the demands of growing populations while prioritizing sustainability, efficiency, and quality of life. This quantum-driven transformation promises to create cities that are smarter, more resilient, and harmonious with their natural and economic environments.

Additionally, quantum computing can optimize land use. By analyzing geographical, environmental, and economic factors, quantum algorithms can recommend the optimal locations for new developments, striking a balance between ecological preservation and urban expansion. This holistic approach reduces urban sprawl and promotes sustainable growth.

ENHANCING IOT NETWORKS WITH QUANTUM COMPUTING

IoT networks form the backbone of smart cities, connecting devices such as traffic lights, sensors, and home appliances to create a cohesive and intelligent system. These systems provide the infrastructure to monitor, control, and optimize various urban functions. However, managing these networks is a complex task, especially as the number of IoT devices grows exponentially, generating vast amounts of data and requiring seamless coordination. Traditional computing methods struggle to handle such complexity, but quantum computing offers a groundbreaking approach to improving data processing, communication, and security across these networks.

DATA PROCESSING AND OPTIMIZATION

IoT devices generate enormous amounts of data that require real-time analysis to be effective. Quantum computing's unmatched processing power enables the simultaneous analysis of multiple data streams, allowing for faster and more accurate decision-making than ever before. For instance, a quantum-powered IoT network could dynamically adjust traffic lights across an entire city to alleviate congestion while optimizing energy distribution in real time. This large-scale coordination is crucial for cities facing population growth and increasing urban density.

Quantum algorithms are also particularly adept at solving complex optimization problems. Delivery services, for instance, can utilize quantum-enhanced IoT networks to determine the most efficient routes for vehicle fleets, thereby reducing fuel consumption and overall delivery times. Likewise, waste management systems could leverage quantum insights to optimize collection schedules and routes, reducing operational costs, minimizing environmental impact, and improving urban sanitation. The ability to address these logistical challenges dynamically transforms how cities manage their resources and services.

REVOLUTIONIZING COMMUNICATION

Communication is the lifeline of IoT networks, enabling devices to share data and collaborate effectively. Quantum computing revolutionizes communication within IoT systems by introducing quantum key distribution (QKD), a method of secure data transmission rooted in the principles of quantum mechanics. Unlike traditional encryption methods, QKD offers unparalleled security; any attempt to intercept the transmission disrupts the quantum state, immediately alerting the system to the breach. This is especially critical for IoT networks that handle sensitive data, including personal information and details about critical infrastructure.

In addition to enhancing security, quantum technologies improve data transmission efficiency in densely populated urban areas. Quantum repeaters and entangled photon systems enable the seamless transmission of data over long distances, significantly reducing latency and boosting the performance of IoT networks. This capability is essential for applications such as real-time surveillance, autonomous vehicles, and emergency response systems, which rely on instantaneous communication.

SECURITY AND RESILIENCE

As IoT networks become increasingly integral to city operations, they become more attractive targets for cyberattacks. Quantum computing addresses this vulnerability by providing robust security measures. In addition to QKD, advanced quantum algorithms can analyze network behavior to detect anomalies and potential threats. These algorithms identify vulnerabilities and neutralize them before they can cause significant damage, ensuring the resilience and reliability of smart city infrastructures.

Furthermore, quantum computing enhances disaster recovery capabilities for IoT networks, offering unprecedented precision and adaptability in crisis scenarios. In natural disasters such as earthquakes, hurricanes, or floods, quantum-powered simulations can process vast amounts of real-time data to accurately model recovery scenarios. These simulations take into account variables such as resource availability, population density, and infrastructure vulnerabilities, enabling effective prioritization of resource allocation. For instance, quantum algorithms can help direct emergency services to the areas of greatest need, ensuring faster and more efficient responses to life-threatening situations.

This proactive approach minimizes downtime for critical city functions, including power grids, transportation systems, and communication networks. During emergencies, the ability to restore essential services swiftly can mean the difference between chaos and order. Quantum computing also aids in predicting secondary impacts, such as the likelihood of aftershocks following an earthquake or potential cascading failures in interconnected systems, enabling preemptive measures to mitigate further damage.

Beyond immediate disaster recovery, quantum computing contributes to long-term resilience by optimizing the planning and maintenance of Internet of Things (IoT) networks. By identifying weak points in infrastructure and simulating stress tests under various disaster scenarios, cities can fortify their systems against future disruptions. This level of preparedness fosters greater public trust as residents see tangible improvements in their city's ability to withstand and recover from crises.

Additionally, quantum technologies can facilitate more robust coordination among different agencies involved in disaster management. By integrating data from diverse sources, such as meteorological systems, health services, and transportation networks, quantum-powered platforms can create a unified response strategy that reduces redundancies and maximizes efficiency. This holistic approach ensures that critical resources, such as medical supplies and emergency shelters, are deployed where they are needed most, thereby reinforcing the reliability and effectiveness of smart city infrastructures. Ultimately, by safeguarding IoT networks and enhancing their disaster recovery capabilities, quantum computing strengthens the backbone of smart cities and paves the way for a future where communities can thrive despite adversity.

APPLICATIONS IN SMART CITIES

Quantum computing and IoT integration are already beginning to redefine urban living. Below are key areas where these technologies intersect to create more innovative, more connected communities:

1. Transportation and Traffic Management

Quantum-enhanced IoT networks can revolutionize urban mobility. Quantum algorithms can optimize traffic flow, reduce congestion, and improve public transportation efficiency by analyzing real-time traffic data. For example, dynamic rerouting of vehicles based on quantum-processed traffic predictions can save commuters time and reduce carbon emissions.

2. Energy Management

Quantum computing facilitates more efficient energy distribution. Smart grids, powered by quantum algorithms, can balance supply and demand in real-time, reducing energy wastage. Quantum simulations can also optimize the integration of renewable energy sources, such as solar and wind power, into urban power systems, thereby promoting sustainability.

3. Healthcare

In a smart city, IoT devices can monitor public health metrics, such as air quality and disease outbreaks. Quantum computing enhances these systems by analyzing data trends and predicting health crises. For instance, during a pandemic, quantum-powered IoT networks could identify hotspots and optimize resource allocation, such as hospital beds or vaccines, thereby enhancing public health.

4. Public Safety

Quantum computing strengthens IoT-based surveillance and

emergency response systems. For instance, quantum algorithms can analyze data from security cameras and sensors to detect unusual activities, enabling quicker responses to potential threats. In disaster management, quantum simulations can model evacuation scenarios to ensure the safety of residents.

5. Urban Planning and Infrastructure Maintenance

Quantum-enhanced IoT networks can monitor the health of urban infrastructure, such as bridges and roads. By analyzing data from embedded sensors, these systems can predict maintenance needs, preventing costly repairs and ensuring public safety. Quantum simulations can also model the impact of new construction projects on existing infrastructure, thereby facilitating more informed planning.

CHALLENGES AND THE PATH FORWARD

Despite its enormous potential, integrating quantum computing with IoT networks in smart cities presents many significant challenges. Quantum hardware, though groundbreaking, remains in its infancy, with persistent issues such as error rates, decoherence, and the sheer complexity of scaling quantum systems to meet urban demands. These limitations currently hinder the widespread adoption of quantum technologies in practical, real-world applications. Additionally, implementing quantum solutions requires substantial investment in cutting-edge infrastructure and the development of a highly skilled workforce capable of managing these advanced systems. Building and maintaining quantum computing facilities involves high costs and specialized equipment, making them accessible only to a limited number of governments and large corporations. For smaller municipalities or less-developed regions, these barriers can seem insurmountable.

Another pressing concern is the integration of quantum systems with existing Internet of Things (IoT) networks. Transitioning from classical computing frameworks to quantum-enhanced systems requires extensive research, testing, and reconfiguration of current infrastructures, which can be time-consuming and costly. The need for seamless compatibility and interoperability adds further complexity to this transition, requiring a coordinated effort between quantum researchers, IoT developers, and city planners.

However, despite these hurdles, progress is accelerating at an impressive pace. Governments and leading technology companies are making unprecedented investments in quantum research, recognizing its potential to transform industries and address global challenges. Tech giants such as IBM, Google, and Microsoft, along with innovative startups, are actively working to overcome the technical difficulties of quantum hardware and software. Simultaneously, educational institutions worldwide are launching specialized programs to train the next generation of quantum scientists, engineers, and technicians, ensuring a steady talent pipeline to support the field's growth.

Moreover, international collaborations are fostering innovation by pooling resources and expertise. Joint ventures between the public and private sectors are helping to fund quantum research while accelerating the development of scalable, user-friendly solutions. These collective efforts are gradually reducing the barriers to quantum adoption, making it more accessible to cities of all sizes and capabilities.

As quantum technologies continue to evolve, we can expect breakthroughs that will address current limitations, paving the way for widespread implementation in smart cities. With sustained investment, innovation, and education, the future of quantum-enhanced urban living is not only possible but inevitable.

ENABLING SMARTER, CONNECTED COMMUNITIES

The convergence of quantum computing and IoT networks heralds a new era of urban living, one in which technological advancements redefine how we interact with and experience our cities. Smarter cities will be more efficient, sustainable, resilient, inclusive, and capable of addressing once-insurmountable challenges. By processing vast datasets and simulating intricate systems, quantum-powered insights unlock innovative solutions to problems ranging from traffic congestion to energy optimization. These transformative technologies promise to enhance the quality of life for millions worldwide by creating urban environments that are adaptable, intelligent, and deeply interconnected.

As we look to the future, the vision of a brilliant city is no longer a distant dream but an achievable reality on the horizon. Quantum computing, coupled with IoT innovations, is the key to unlocking this potential by delivering unprecedented data analysis, communication, and decision-making capabilities. Imagine a world where urban systems not only anticipate the needs of their inhabitants but also evolve in real time to meet them. Traffic systems that dynamically adjust to reduce congestion, energy grids that optimize the use of renewable energy, and healthcare systems that predict and prevent disease outbreaks are just a glimpse of what is possible.

Moreover, integrating quantum technologies fosters a greater sense of inclusivity and resilience. These advancements empower cities to address socioeconomic disparities by optimizing resource allocation and improving access to essential services. For instance, housing projects can be planned with unparalleled efficiency to meet the needs of growing populations, while transportation systems can be designed to provide equitable access to all residents. The possibilities also extend to environmental sustainability, with quantum computing driving innovations in waste reduction, renewable energy integration, and pollution monitoring.

The journey toward realizing these quantum-enhanced smart cities has only just begun, but the foundations are being laid with each technological breakthrough and investment in research. The dream of a quantum-powered urban future becomes more tangible as the public and private sectors collaborate to accelerate progress. The possibilities, as boundless as the imagination, invite us to rethink the fabric of urban life and embrace a future where cities are not only places to live but ecosystems of innovation, sustainability, and human potential.

PART 5

PREPARING FOR THE QUANTUM FUTURE

CHAPTER 13

ETHICAL AND SOCIETAL IMPLICATIONS

Quantum computing represents a monumental leap in technological capability, promising to revolutionize industries and solve complex problems at an unprecedented scale. This advancement ushers in opportunities to redefine how we approach issues like medicine, environmental science, and artificial intelligence. However, with great power comes great responsibility, as the disruptive potential of quantum technologies also poses significant ethical and societal challenges. As we stand on the brink of the quantum era, these concerns are no longer theoretical but immediate and pressing.

The transformative potential of quantum computing necessitates addressing two primary areas of concern. First, data privacy in a post-quantum world faces unprecedented risks. Our current encryption methods, designed to withstand classical computing capabilities, are no match for quantum algorithms that can crack them in moments. Second, quantum disruption raises complex ethical questions about fairness, equity, and the responsible use of these groundbreaking tools. What happens when quantum advantages fall disproportionately to a select few? How can we ensure this powerful technology benefits humanity as a whole? This chapter examines these critical areas to provide a comprehensive understanding of the key issues at stake. It will explore the vulnerabilities exposed by quantum advancements, delve into the moral complexities of its applications, and offer a roadmap for navigating this uncharted territory responsibly.

DATA PRIVACY CONCERNS IN A POST-QUANTUM WORLD

The advent of quantum computing poses a significant challenge to current data security measures. Traditional encryption methods, such as RSA and elliptic curve cryptography (ECC), rely on the computational difficulty of factoring large numbers or solving discrete logarithm problems. These tasks are impossible for classical computers to solve within a reasonable timeframe, making them the backbone of modern cybersecurity.

Quantum computers, however, threaten to upend this security framework in once inconceivable ways. Algorithms like Shor's algorithm, designed specifically for quantum systems, can solve these problems exponentially faster than classical algorithms. This means nothing short of revolutionary—encrypted communications, financial transactions, and sensitive data that underpin our global digital infrastructure could be rendered vulnerable. What takes classical systems thousands or even millions of years to compute for a sufficiently powerful quantum computer could be achieved in mere hours or minutes. This would effectively dismantle the protective walls that currently shield sensitive information from prying eyes.

Governments, corporations, and individuals face unprecedented risks in a post-quantum world. Consider the implications of state-sponsored actors or sophisticated cybercriminal groups gaining access to quantum decryption tools. Entire databases of confidential information—ranging from military strategies and government communications to the intellectual property of leading tech firms—could be compromised. Furthermore, personal data, such as medical records and financial histories, stored under today's encryption standards could be exposed retroactively, revealing a wealth of information long after it was thought to be secure. The cascading effects of such breaches could disrupt economies, destabilize societies, and undermine public trust in digital systems.

Moreover, this quantum threat introduces a profound sense of urgency for cybersecurity professionals and policymakers. We risk entering an era without private information if we do not take preemptive steps to safeguard data. Individuals could not rely on encrypted messaging platforms, businesses might have their trade secrets exposed, and governments could find themselves at a strategic disadvantage. The stakes are exceptionally high for industries like healthcare and finance, where the confidentiality and integrity of data are critical not only to operations but also to public trust.

The ripple effects of these challenges extend beyond technical vulnerabilities. They provoke broader societal questions about how we value privacy in the digital age. Will the advent of quantum computing redefine our expectations of what it means to have secure communications? How will public and private institutions rebuild trust in systems that were once considered unbreakable? These questions demand immediate and sustained attention as the quantum future rapidly approaches.

MITIGATING THE RISKS

To address these challenges, researchers and policymakers are developing quantum-safe cryptographic protocols that promise resilience against the computational power of quantum systems. These post-quantum algorithms are designed to resist attacks from classical and quantum computers. Examples include lattice-based cryptography, hash-based signatures, and code-based encryption. These approaches leverage mathematical problems that remain computationally infeasible even for quantum computers. While promising, transitioning to quantum-safe cryptography globally is a monumental task. This shift requires an overhaul of existing infrastructure, significant investment in research, and training for cybersecurity professionals to handle new cryptographic standards. It also demands robust testing to ensure these algorithms can withstand future threats.

Organizations must adopt a comprehensive and proactive approach to safeguarding data. Techniques such as hybrid encryption, which combines classical and quantum-safe methods, offer interim solutions while the technology matures. Such methods ensure redundancy, preserving security even as systems transition. Furthermore, developing scalable quantum-resistant protocols will require international cooperation to establish global standards. Governments and international bodies must prioritize collaborative efforts to create regulations that govern quantum-era cybersecurity. This includes technical solutions and frameworks to address the ethical implications of data security in the quantum age.

Public awareness campaigns are equally vital in preparing society for these changes. Individuals, businesses, and institutions must understand the significance of quantum computing and the risks it poses to data privacy. Educational initiatives can bridge the knowledge gap, fostering a culture of vigilance and adaptability. Cybersecurity workshops, online courses, and informational resources must be widely accessible to empower people to take preventive measures.

In addition to technical and educational efforts, strategic investments in research and development are crucial. Public-private partnerships can accelerate the development of innovative cryptographic solutions while ensuring deployment across diverse industries. Encouraging collaboration between academic institutions, tech companies, and government agencies will help harmonize efforts and streamline the transition to quantum-safe environments. By fostering an ecosystem of shared knowledge and mutual support, stakeholders can collectively mitigate the risks of quantum computing, paving the way for a more secure digital future.

ETHICAL DILEMMAS OF QUANTUM DISRUPTION

Beyond data privacy, quantum computing raises broader ethical questions about its potential to disrupt industries, economies, and societies. While the benefits of quantum technologies are immense, their rapid adoption could exacerbate existing inequalities and create unforeseen challenges. The unprecedented computational power of quantum systems could lead to significant shifts in competitive advantage, favoring entities with early access to these technologies. This advantage may result in monopolistic control over critical industries, including finance, healthcare, and artificial intelligence, creating a landscape where smaller players struggle to compete effectively.

Moreover, the transformative nature of quantum computing could intensify global economic disparities. Wealthy nations and corporations with the resources to develop and deploy quantum technologies may gain a significant advantage, leaving less-developed regions further behind. This technological gap could undermine efforts toward global equity, as underprivileged communities are unable to harness quantum innovations for their benefit.

The societal impact of quantum disruption extends to individual livelihoods as well. Jobs in logistics, pharmaceuticals, and cybersecurity industries could face significant upheaval as quantum algorithms streamline operations and solve problems at unprecedented scales. While some roles will evolve or adapt to the quantum era, others may disappear entirely, requiring comprehensive workforce strategies to mitigate displacement and promote equitable opportunities in emerging quantum sectors.

In addition to these economic and labor-related concerns, quantum disruption challenges existing regulatory and ethical frameworks. Quantum technologies' dual-use potential—their ability to produce both beneficial and harmful outcomes—necessitates careful oversight. Without adequate safeguards, quantum capabilities could be weaponized to facilitate large-scale cyberattacks, invasive surveillance, or the development of highly efficient destructive technologies. Establishing international consensus on the ethical boundaries of quantum research and deployment is imperative to prevent misuse and ensure this transformative technology serves the greater good.

ECONOMIC DISPARITIES

Quantum computing has the potential to significantly widen the gap between technologically advanced nations and those lacking access to cutting-edge research, funding, and resources.

Countries and corporations that dominate the quantum landscape may secure significant advantages in key fields such as finance, healthcare, and defense. Such dominance could translate into unparalleled strategic and economic control, perpetuating inequalities and creating a global power imbalance. Developing nations, already struggling with limited technological access, may find themselves increasingly marginalized in the global economy, as they compete with quantum-enabled industries in wealthier regions.

Furthermore, the disparity extends beyond nations to corporations within technologically advanced societies. Companies with the capital to invest in quantum technologies are likely to outpace their competitors, consolidating market share and creating monopolistic dynamics in emerging quantum-powered industries. This concentration of power risks stifling innovation and limiting opportunities for smaller enterprises and startups.

Proactive measures are crucial in addressing these disparities. One of the most effective ways to democratize quantum computing is through international collaborations that pool resources and expertise across borders. Developing global partnerships in quantum research can ensure more equitable access to this transformative technology. Open-source quantum software initiatives, such as those spearheaded by academic institutions and nonprofit organizations, can also play a pivotal role by making advanced tools and algorithms available to a broader audience. Such efforts enable smaller nations and organizations to engage with quantum technology without prohibitive costs.

Educational initiatives tailored to the unique needs of under-resourced regions can further bridge the gap. By building a skilled quantum workforce through accessible training programs, nations can create opportunities to participate in the quantum revolution. Funding mechanisms, such as grants and subsidies for quantum research in underdeveloped areas, can encourage investment and foster local innovation. Encouraging multinational corporations to support quantum research and development across diverse geographies can also distribute the benefits of quantum computing more evenly, ensuring that no single entity monopolizes its transformative potential.

Finally, creating a framework for equitable technology-sharing agreements can ensure that quantum breakthroughs benefit humanity. For example, global organizations can establish initiatives to share the outcomes of quantum advancements in critical areas such as healthcare and climate solutions with all nations, regardless of their technological capacity. By fostering an inclusive approach, the global community can work to mitigate the risks of quantum-induced disparities and unlock its full potential for collective progress.

IMPACT ON EMPLOYMENT

The automation potential of quantum computing may disrupt labor markets, particularly in industries reliant on optimization and complex problem-solving. For example, quantum algorithms could streamline supply chain management, rendering specific logistical roles obsolete. Similarly, advancements in drug discovery powered by quantum simulations might reduce the need for traditional research roles.

While quantum computing will create new opportunities, such as careers in quantum software development, hardware engineering, and ethical oversight, governments and organizations must prepare for workforce transitions. Policies focused on reskilling and upskilling workers, as well as fostering innovation in complementary industries, will be critical to minimizing economic dislocation.

ETHICAL USE CASES

The dual-use nature of quantum computing technologies raises questions about their ethical application. On one hand, quantum algorithms can solve critical problems in healthcare, energy, and climate science. On the other hand, they could be weaponized for malicious purposes, such as enhancing cyberattacks or developing advanced surveillance systems.

Ethical frameworks and guidelines are needed to ensure quantum technologies are used responsibly. Stakeholders, including governments, corporations, and academic institutions, must collaborate to establish quantum research and application norms. Transparency in quantum development, combined with robust oversight mechanisms, can prevent misuse and foster public trust.

ENVIRONMENTAL CONSIDERATIONS

Quantum computers, particularly those relying on superconducting qubits, require significant energy to maintain ultra-cold operating environments. As the technology scales, its environmental impact must be carefully managed. To reduce the carbon footprint of quantum computing, sustainable energy sources, such as solar or wind power, should be prioritized.

Moreover, the broader implications of quantum advancements on environmental policy should be considered. For instance, quantum simulations enable breakthroughs in renewable energy technologies but also lead to the exploitation of natural resources for quantum materials. Balancing innovation with environmental stewardship will be a critical ethical challenge.

THE ROLE OF EDUCATION AND PUBLIC AWARENESS

Empowering society to navigate the ethical and societal implications of quantum computing begins with education. Universities, governments, and private organizations must invest in programs that teach the technical aspects of quantum computing and its broader impacts. These educational initiatives must be accessible and inclusive, reaching diverse populations through online courses, workshops, and community outreach programs. By doing so, society can build a foundation of knowledge that encourages widespread engagement with quantum technologies. Public awareness campaigns can further demystify quantum technology, illustrating its benefits and risks through relatable examples and accessible language. Such efforts are instrumental in fostering a well-informed public capable of engaging in meaningful discussions about the ethical use of quantum innovations.

Additionally, interdisciplinary collaboration between technologists, ethicists, policymakers, and the public will be vital in addressing the complexities of quantum ethics. Forums, think tanks, and advisory panels focused on quantum ethics can provide platforms for diverse perspectives, enabling experts and stakeholders to identify potential risks and devise proactive strategies. These collaborative efforts can create holistic solutions that balance innovation with ethical considerations by integrating viewpoints from various disciplines. Furthermore, regular public consultations and transparent communication from industry leaders and researchers can help strengthen trust and ensure that quantum technologies are developed and deployed with a focus on societal well-being.

Beyond formal collaboration, grassroots movements and citizen science initiatives can also shape the ethical trajectory of quantum computing. These community-driven efforts can amplify the voices of underrepresented individuals, ensuring that the moral discourse surrounding quantum technologies reflects the interests and concerns of a broad spectrum of society. By fostering an environment of inclusivity and shared responsibility, society can better navigate the profound changes that quantum computing promises to bring.

As we enter the quantum era, the ethical and societal implications of this transformative technology cannot be overlooked. From safeguarding data privacy to addressing economic disparities and ensuring responsible use, quantum computing presents challenges and opportunities. This technology can redefine entire industries, reshape economies, and transform the functions of societies. However, its potential comes with immense risks that demand immediate and thoughtful action.

By adopting a proactive and collaborative approach, we can mitigate the risks associated with quantum computing and harness its potential for the greater good. For instance, fostering international cooperation can ensure equitable access to quantum technologies while ethical guidelines safeguard against misuse. Public education campaigns, too, are critical to demystifying quantum computing, enabling societies to engage in informed discussions about its impact.

The choices we make today will shape the trajectory of this powerful technology and its impact on future generations. By building robust frameworks for ethical oversight, investing in quantum-safe infrastructure, and promoting inclusive policies, we can create a future where quantum computing serves as a force for collective progress rather than division. We must ensure that this transformative era leads to a more secure, equitable, and innovative world.

CHAPTER 14

GETTING QUANTUM-READY

Quantum computing, a field once limited to theoretical physics and academia, is becoming increasingly relevant across industries. Its transformative potential drives innovation across technology, healthcare, finance, logistics, and beyond. As quantum computing transitions from research labs to real-world applications, it opens many opportunities for individuals and businesses. This emerging field offers high-impact career pathways and promises to reshape global industries. As quantum advancements accelerate, understanding their implications and preparing for their adoption has never been more crucial. This chapter explores career pathways in quantum computing, the tools and platforms available for learning, and actionable steps for preparing for the quantum revolution. By studying these areas, we aim to provide a roadmap for anyone seeking to participate in this groundbreaking journey.

CAREER OPPORTUNITIES IN QUANTUM COMPUTING

The quantum computing ecosystem is expanding rapidly, creating diverse and increasingly specialized career opportunities. Roles in this field encompass research, development, engineering, business strategy, and education, with new interdisciplinary opportunities emerging as the field continues to evolve. The demand for expertise in quantum computing is driving innovation across sectors, prompting industries to invest heavily in cultivating talent. Here are some key career paths, along with insights into their growing significance:

QUANTUM SOFTWARE DEVELOPER

Quantum software developers create algorithms and applications optimized for quantum computers. Familiarity with quantum programming languages like Qiskit, Cirq, or Q# is essential. Developers often work on problems in cryptography, optimization, and machine learning. These professionals are vital for translating quantum theories into practical, real-world solutions that industries can adopt.

QUANTUM HARDWARE ENGINEER

Hardware engineers design and build the physical systems underlying quantum computers. This role requires expertise in materials science, electrical engineering, and quantum mechanics. Engineers tackle challenges such as improving qubit stability and minimizing noise. They work closely with research teams to bridge the gap between theoretical designs and functional prototypes, driving the evolution of quantum hardware.

QUANTUM RESEARCH SCIENTIST

Research scientists focus on advancing the theoretical and practical understanding of quantum computing. This role often requires a Ph.D. in quantum physics, computer science, or a related field. Scientists work on new quantum algorithms, error correction methods, or novel hardware architectures. Their contributions lay the groundwork for breakthroughs that can redefine industries and open new frontiers of knowledge.

BUSINESS STRATEGIST OR CONSULTANT

As businesses adopt quantum technologies, strategists must identify use cases, evaluate the return on investment (ROI), and develop implementation plans. This role bridges the gap between technical teams and business leaders. They help organizations navigate the complexities of quantum computing, ensuring a seamless transition and alignment with strategic goals.

EDUCATORS AND TRAINERS

The demand for quantum literacy is growing. Educators and trainers play a critical role in upskilling professionals and students.

This path may involve teaching at universities, creating online courses, or leading workshops. These educators often collaborate with industry leaders to ensure their curricula remain relevant and forward-thinking.

INTERDISCIPLINARY ROLES

Quantum computing intersects with fields like biology, finance, and logistics. Professionals with expertise in these domains can collaborate with quantum specialists to solve industry-specific problems. For example, a biologist with quantum knowledge might contribute to drug discovery, while a financial analyst could use quantum tools for risk management.

TOOLS FOR LEARNING QUANTUM COMPUTING

As the quantum ecosystem grows, so does the availability of learning resources. This expansion reflects a collective effort by academia, industry leaders, and open-source communities to make quantum computing more accessible than ever before. Learners now have an unprecedented range of platforms, tools, and materials to explore this complex field, allowing them to build skills at their own pace and according to their specific interests. These resources offer technical guidance and foster an environment of collaboration and innovation, which is critical for mastering such a rapidly evolving discipline. The following platforms and tools are instrumental for gaining hands-on experience and unlocking the potential of quantum computing:

QISKIT (IBM)

Qiskit is an open-source quantum programming framework developed by IBM. It provides tools for building and running quantum algorithms on simulators and IBM's quantum hardware. Qiskit's tutorials and documentation make it an excellent starting point for beginners. Its community support ensures learners can access help and guidance at every step. Qiskit enables users to explore a wide range of functionalities, from basic quantum gates to complex machine-learning models, by offering a modular and extensible structure. The IBM Quantum Experience platform complements Qiskit, providing free access to real quantum hardware for experimentation.

GOOGLE CIRQ

Cirq, developed by Google, is a Python library for designing and simulating quantum circuits. It focuses on near-term quantum devices and is particularly useful for exploring variational algorithms and hybrid quantum-classical workflows. Google Cirq also integrates with other tools, allowing developers to create complex simulations with relative ease. By offering detailed tutorials and active community forums, Cirq enables users to delve deeply into quantum experiments at all levels. Its compatibility with TensorFlow Quantum further strengthens its appeal for developers interested in quantum-enhanced AI applications.

AWS BRACKET

Amazon's AWS Bracket offers access to quantum computers from multiple providers, including IonQ and Rigetti. It supports a range of quantum programming frameworks, making it a versatile tool for developers. AWS Braket's cloud-based model makes high-quality quantum computing resources accessible to researchers and businesses worldwide. The platform's integration with the Amazon Web Services ecosystem provides scalability and additional computing power for complex simulations. Moreover, AWS Braket's pricing model offers flexibility, enabling cost-effective experimentation for both small teams and large enterprises.

MICROSOFT AZURE QUANTUM

Azure Quantum is a cloud platform that provides access to quantum hardware and simulators. It integrates with Microsoft's Quantum Development Kit (QDK), which includes the Q# programming language. With robust documentation and enterprise-level support, Azure Quantum is well-suited for businesses exploring quantum applications. Its emphasis on hybrid workflows enables users to blend classical and quantum solutions seamlessly. Microsoft's partnerships with quantum hardware providers also ensure a diverse range of tools and capabilities, catering to the varied research needs of its users.

OTHER NOTABLE TOOLS

- QuTiP (Quantum Toolbox in Python): Ideal for simulating quantum systems. It allows researchers to experiment with dynamic models and test quantum hypotheses. The tool's extensive library of built-in functions simplifies the simulation of open quantum systems, making it a favorite among academics and researchers.
- PENNYLANE: Focuses on quantum machine learning and hybrid quantum-classical algorithms. This tool is perfect for developers interested in the intersection of AI and quantum technologies. Its compatibility with significant machine learning frameworks, such as PyTorch and TensorFlow, makes it an invaluable asset for AI researchers.
- QUANTUM DEVELOPMENT ENVIRONMENTS: Platforms like Xanadu and D-Wave Leap provide specialized environments for unique quantum tasks. These platforms often include dedicated resources for tackling specific quantum problems, such as optimization and annealing, providing tailored solutions to address industry challenges.

PREPARING FOR THE QUANTUM REVOLUTION

The quantum era is not just for scientists and developers; it is a transformative phase that calls for proactive engagement from people across all sectors. For individuals, this means cultivating a mindset of adaptability and lifelong learning to navigate the complexities of quantum technology. Whether through formal education or hands-on experimentation, individuals can harness the power of quantum computing to unlock new opportunities in their careers and personal projects. Similarly, businesses must align their strategies with the unique demands of quantum innovation, ensuring they stay ahead of the curve. This requires adopting cutting-edge tools and fostering a culture of continuous improvement and collaboration. Governments and educational institutions play a pivotal role in this evolution, creating policies, programs, and funding opportunities that democratize access to quantum technologies. Together, individuals, businesses, and institutions can build an ecosystem where the challenges of quantum computing are met with ingenuity and resilience, laying the groundwork for a future of unprecedented possibilities.

1. INDIVIDUALS: BUILDING QUANTUM SKILLS

- Learn the Basics: Start with foundational concepts like superposition, entanglement, and quantum gates. Coursera, edX, and YouTube are excellent online resources for courses and tutorials. These platforms often feature beginner-friendly content tailored to those without prior knowledge of quantum mechanics. Additionally, interactive simulations and visual aids can simplify abstract concepts, making them more accessible to new learners.
- Practice Programming: Experiment with quantum programming frameworks like Qiskit or Cirq. Many platforms

offer free access to simulators and beginner-friendly exercises. Hands-on practice helps bridge the gap between theoretical understanding and practical application. By solving small-scale quantum problems, learners can build confidence and gain a deeper appreciation for the technology's potential.

- Join Communities: Engage with quantum computing communities on forums like Stack Overflow, Reddit's quantum computing, and Slack groups. Networking with professionals can provide insights and opportunities. These communities often host events, hackathons, and webinars to promote collaboration and learning. Building relationships within these networks can lead to mentorship opportunities and access to cutting-edge projects.
- Pursue Formal Education: For those aiming for advanced roles, pursuing a degree or certification in quantum computing, computer science, or related fields is beneficial. Universities worldwide are introducing specialized programs to meet the growing demand for quantum expertise. Dual-degree options and interdisciplinary programs further enhance career prospects, blending quantum skills with domain-specific knowledge.

2. BUSINESSES: INTEGRATING QUANTUM STRATEGIES

- Understand Use Cases: Identify areas where quantum computing can address current limitations—for example, optimization problems in logistics or risk analysis in finance. Collaborating with quantum experts can help pinpoint high-impact applications. Developing an internal task force dedicated to exploring quantum opportunities can provide a strategic advantage.

- Collaborate with Experts: Partner with quantum research institutions or hire consultants to evaluate your organization's quantum readiness. These collaborations can provide valuable insights into navigating this emerging field. Joint research initiatives and pilot programs can also serve as low-risk entry points for businesses.
- Invest in Training: Upskill employees in quantum computing basics to foster internal expertise. Training programs from providers like IBM and Microsoft can help. By cultivating internal talent, businesses can remain competitive in a quantum-driven market. Hosting internal workshops or leveraging e-learning platforms can democratize access to quantum education across teams.
- Start Small: Experiment with quantum computing by tackling proof-of-concept projects. Cloud platforms like AWS Bracket or Azure Quantum can be used to access quantum resources without significant investment. Such initiatives allow organizations to explore quantum technologies without overcommitting resources. Gradual scaling based on pilot outcomes can optimize resource allocation and maximize impact.

3. GOVERNMENTS AND INSTITUTIONS

Governments and academic institutions also play a vital role in quantum readiness. Policies promoting quantum research, funding initiatives, and collaboration between academia and industry are critical. Governments can create strategic funding models that incentivize research and development in quantum technologies. For instance, public-private partnerships could ensure the translation of theoretical advancements into practical applications, bridging the gap between innovation and industry needs. National quantum strategies that include international collaborations can elevate a country's leadership position in this transformative field. Additionally, robust intellectual property frameworks and global agreements can protect and accelerate the exchange of ideas in the quantum domain.

Educational institutions can develop specialized programs to cultivate a quantum-literate workforce. Beyond traditional degree programs, short-term boot camps, certifications, and online modules targeting professionals from non-quantum backgrounds can make quantum education more inclusive. Universities might collaborate with industry leaders to design curricula that directly address practical challenges, ensuring graduates are prepared for the demands of a rapidly evolving quantum landscape. Outreach programs targeting high school students can inspire early interest in quantum technologies, fostering a pipeline of future talent.

CHALLENGES AND HOW TO OVERCOME THEM

Preparing for quantum computing comes with challenges:

- Steep Learning Curve: Quantum computing concepts can be abstract and counterintuitive. Starting with simplified resources and gradually progressing can help. Mentorship and structured programs, such as immersive workshops or mentorship circles led by experienced professionals, can also accelerate learning. Leveraging gamified learning tools or interactive simulations can make quantum concepts more engaging and less intimidating.
- Limited Access to Hardware: Quantum computers are still scarce and expensive. Cloud-based simulators and platforms offer a practical alternative for most learners. Universities and research institutions can also form consortiums to share quantum resources, ensuring broader access. Programs that subsidize quantum computing time for students and researchers could democratize access to these cutting-edge technologies, enabling more inclusive innovation.
- Uncertain ROI: Businesses may hesitate to invest in a technology still in its infancy. Conducting small-scale experiments and staying informed about advancements can mitigate risks. Organizations can adopt a phased approach, starting with low-cost pilot projects to gradually build expertise and experience. Collaborating with quantum consultants or industry pioneers can provide actionable insights into aligning quantum applications with long-term business goals.

A ROADMAP FOR THE FUTURE

The quantum revolution is a journey, not a destination. Continuous learning and staying adaptable are key for individuals. As the quantum field evolves, the ability to learn new tools and adapt to emerging paradigms will distinguish leaders from followers. Individuals can start by exploring foundational concepts, but must also embrace a mindset of lifelong learning. This may include attending specialized workshops, enrolling in academic programs, or participating in quantum-focused online communities.

Fostering a culture of innovation and collaboration will ensure business success in a world driven by quantum technology. Innovation in this context refers to adopting cutting-edge technologies and empowering employees to experiment and learn from their failures. Collaboration, both internally across departments and externally with quantum research institutions or tech providers, is crucial. Companies that build interdisciplinary teams combining quantum expertise with domain knowledge will be well-positioned to solve unique challenges and seize new opportunities.

Governments and institutions must continue to provide the infrastructure and support necessary to sustain quantum advancements. This includes funding for research and development, as well as crafting policies that promote collaboration among academia, industry, and government. Building international alliances can accelerate progress, ensuring that countries and organizations benefit collectively from breakthroughs in quantum technology.

Institutions should prioritize creating quantum literacy programs for students and professionals, as this knowledge will become increasingly indispensable shortly.

By embracing quantum computing today, we prepare ourselves for a future where the impossible becomes possible. Quantum technology promises to redefine industries, enhance scientific discovery, and tackle challenges previously considered unsolvable. Whether you're a curious learner, an aspiring professional, or a forward-thinking business leader, the time to get quantum-ready is now. This journey is not merely about technological transformation—it's about reshaping humanity's approach to problem-solving, innovation, and collaboration. As we stand on the cusp of this quantum leap, we must all play our part in building a future where the vast potential of quantum computing can be fully realized.

CONCLUSION

RECAP OF QUANTUM COMPUTING'S TRANSFORMATIVE POTENTIAL AND CHALLENGES

Quantum computing represents one of the most groundbreaking technological advancements of our era. Unlike classical computers that rely on bits, quantum computers harness the unique properties of qubits—such as superposition and entanglement—to process information in ways once unimaginable. By leveraging these principles, quantum computers can perform calculations that would otherwise be intractable, enabling them to tackle some of humanity's most complex problems. These capabilities redefine the scope of computation, enabling new approaches to solving issues across science, technology, and society.

These unique properties of quantum systems empower researchers and developers to conceptualize and execute solutions that classical computers could not approach, even with centuries of processing power. For example, simulating the complex behavior of molecules for drug discovery or understanding the intricate dynamics of financial markets are tasks where quantum computing excels. This paradigm shift opens the door to innovations previously thought impossible, redefining the boundaries of what technology can achieve. Furthermore, the implications extend beyond technical applications, influencing global industries, governance, and the pursuit of scientific discovery. Quantum computing is a testament to the boundless potential of human ingenuity, marking a transformative era for technology and society.

Throughout this book, we've explored the vast potential of quantum computing to revolutionize fields critical to humanity's progress. In artificial intelligence, quantum acceleration enables machines to learn and adapt at unprecedented speeds, transforming industries from healthcare to finance. In cryptography, quantum computers challenge traditional encryption systems while inspiring the development of quantum-safe alternatives, redefining the cybersecurity landscape. The ability to simulate molecular interactions with extraordinary precision accelerates breakthroughs in drug discovery, renewable energy storage, and even climate modeling, addressing some of the world's most pressing challenges.

Quantum computing can also optimize complex systems, such as supply chains, transportation networks, and financial markets, providing insights that classical models struggle to uncover. Imagine logistics networks so finely tuned that they minimize waste, reduce costs, and deliver goods faster than ever, creating a ripple effect across global economies. In transportation, quantum algorithms can help design traffic systems that adapt in real-time, significantly reducing congestion and emissions and paving the way for greener cities. Similarly, quantum computing can enhance fraud detection, risk assessment, and portfolio optimization in financial markets, ensuring excellent stability and efficiency in economic systems.

Furthermore, its role in scientific research—from exploring the fundamental nature of particles to unraveling the universe's mysteries—highlights its transformative impact on human understanding. By simulating molecular interactions with extraordinary precision, quantum computers enable researchers to design new materials with unprecedented properties, such as superconductors that could revolutionize energy transmission. They also provide insights into fundamental questions about the nature of reality itself, offering a deeper understanding of the quantum world and its implications for our universe. In essence, quantum computing is not just a tool but a gateway to new frontiers of human knowledge and capability.

Yet, this transformative potential is accompanied by significant challenges. Technical hurdles, including noise, decoherence, and the need for advanced error correction, currently limit the capabilities of quantum computers. The physical demands of quantum hardware—cryogenic cooling systems and specialized materials—underscore the immense resources required to scale these systems. Additionally, the disruptive nature of quantum technology raises profound ethical and societal concerns. How do we protect sensitive data in a post-quantum world? How do we ensure equitable access to quantum breakthroughs, avoiding a technological divide? Addressing these challenges demands a collaborative, multidisciplinary approach from governments, industries, and academia worldwide.

Yet, this transformative potential comes with significant challenges. Noise, decoherence, and error correction issues currently limit the capabilities of quantum computers. Building and maintaining quantum hardware requires substantial resources and expertise, and scalable, fault-tolerant quantum computers remain a distant goal. Moreover, the disruptive nature of quantum computing raises ethical and societal questions, from data privacy concerns in a post-quantum world to the equitable distribution of quantum-enabled advancements. These challenges are substantial, but they are not insurmountable.

A HOPEFUL VISION FOR A QUANTUM-ENHANCED FUTURE

Despite the hurdles, the future of quantum computing is bright and full of promise. Imagine a world where personalized medicine becomes the norm, thanks to quantum-powered simulations that tailor treatments to an individual's unique genetic makeup. Quantum algorithms could analyze vast amounts of genetic and molecular data in seconds, enabling doctors to predict diseases before they manifest and devise precise interventions. Envision cities that operate more efficiently, with quantum algorithms seamlessly managing traffic systems to eliminate congestion, optimizing energy grids to reduce waste, and streamlining urban planning to accommodate population growth in a sustainable manner. Imagine interconnected smart cities where quantum computing enables real-time decision-making for more effective resource allocation and enhanced environmental protection.

Consider the possibilities of advancing clean energy solutions, such as designing new battery materials that store more energy and charge more quickly, or optimizing solar panel efficiency to harness the maximum energy from the sun. Revolutionary materials, designed atom by atom using quantum simulations, could lead to innovations such as ultra-lightweight, ultra-strong alloys for transportation or superconductors that enable lossless energy transmission. Beyond material science, quantum computing could unlock new scientific phenomena, revealing insights into the nature of dark matter or the universe's origins, fundamentally transforming our understanding of existence, and propelling humanity into a new age of discovery and innovation.

As we stand on the cusp of the quantum era, collaboration and innovation will be key. Governments, academic institutions, and private organizations invest heavily in quantum research and development. These collective efforts are driving the field forward, fostering an ecosystem that enables quantum technology to thrive. While the road ahead is long, each milestone brings us closer to realizing the full potential of this transformative technology.

Quantum computing is not just a subject for scientists and technologists; it is a revolution that will impact every aspect of society. Understanding its principles and implications is crucial for navigating the changes it will bring. Whether you are a student exploring career opportunities, a business leader strategizing for the future, or a curious mind eager to learn, now is the time to engage with the quantum revolution.

Take the first step by continuing to educate yourself. Explore quantum programming tools like Qiskit, Google Cirq, and AWS Bracket. Dive deeper into the latest research papers and attend seminars that demystify the quantum world. Follow developments from industry leaders, such as IBM and Google, as well as emerging startups that are pushing the boundaries of innovation. Advocate for ethical considerations and equitable access to quantum technologies to ensure no one is left behind in this transformative journey. By engaging actively, you contribute to shaping a future that leverages quantum computing for the collective good.

The quantum revolution is not just an opportunity for innovation and progress; it is a call to action for all of humanity. Quantum computing challenges us to think beyond the boundaries of classical paradigms, offering a glimpse into a world of new possibilities. It invites us to collaborate across disciplines, reimagine existing systems, and address some of the most pressing challenges of our time with unprecedented tools and methods. This revolution leads students to exciting careers at the forefront of technology.

For businesses, it presents an opportunity to unlock unprecedented efficiencies and benefits. For the curious, it presents a chance to witness and understand the dawn of a new era in computation and science.

By embracing this technology with curiosity, responsibility, and a forward-thinking mindset, we can ensure that the quantum leap leads to a brighter, more connected, and more innovative future. Let us approach this journey with purpose and optimism, recognizing that our actions today will shape the quantum landscape of tomorrow. The journey begins with understanding, and the destination is limitless. Together, we can embrace the quantum revolution and transform potential into reality.

GLOSSARY

AIRBUS: Optimizing aircraft design. Airbus uses quantum computing to optimize aircraft design and performance in the aerospace industry.

Algorithmic Trading: Quantum algorithms redefine the landscape of algorithmic trading by processing vast amounts of market data at unparalleled speeds.

BASF: Accelerating chemical research. Chemical giant BASF is leveraging quantum computing to accelerate research and development in material science.

BUILDING QUANTUM SKILLS: Learn the basics: Start with foundational concepts like superposition, entanglement, and quantum gates.

Collaborate with Experts: Partner with quantum research institutions or hire consultants to evaluate your organization's quantum readiness.

Computing with Quantum Cats: From Colossus to Qubits.

CURRENT LEADERS (GOOGLE, MICROSOFT, AND INNOVATIVE STARTUPS) The quantum computing race has drawn some of the world's most influential companies.

Customer Insights: Quantum data analysis capabilities extend beyond operational efficiencies to enhance customer experiences.

Dancing with Qubits: how quantum computing works and how it can change the world.

Decoherence: The process where qubits lose their quantum state due to environmental interference, limiting computational reliability.

DHL and Quantum Optimization: The global logistics giant DHL has been at the forefront of experimenting with quantum computing to optimize its supply chain operations.

Dreams versus Reality: Plenary debate session on quantum computing.

Driven Routing Systems: FedEx, a global leader in package delivery, has significantly invested in exploring quantum computing to improve its routing systems.

ENTANGLEMENT: The power of correlation. Entanglement is one of the most extraordinary and unique properties of quantum computing.

Error Correction: Techniques in quantum computing to manage errors like bit-flip and phase-flip without collapsing quantum states.

EVOLUTION OF QUANTUM COMPUTING: From Feynman to the present. The concept of quantum computing emerged from the visionary ideas of physicist Richard Feynman in the early 1980s.

Five Starter Problems: Solving quadratic unconstrained binary optimization models on quantum computers.

Grover's Algorithm: A quantum search algorithm offering quadratic speedup for unstructured search problems.

Hybrid Quantum-Classical Systems: Combining classical and quantum computing for practical applications while quantum hardware matures.

INTEGRATING QUANTUM STRATEGIES: Understand Use Cases: Identify areas where quantum computing can address current limitations.

Invest in Training: Upskill employees in quantum computing basics to foster internal expertise.

Join Communities: Engage with quantum computing communities on forums like Stack Overflow, Reddit's quantum computing, and Slack groups.

JP MORGAN CHASE (FINANCIAL MODELING) JP Morgan Chase is exploring quantum computing in the financial sector to improve risk analysis and portfolio optimization.

Limited Access to Hardware: Quantum computers are

still scarce and expensive.

Logistics Optimization: Using quantum systems to streamline supply chain management and routing in industries like transportation.

Material Discovery: Leveraging quantum simulations to design novel materials with desired properties.

Microsoft Azure Quantum: A platform integrating quantum computing with cloud services for broad accessibility.

Noise: Unwanted disturbances in quantum systems caused by environmental factors, introducing errors in computations.

Optimization in Smart Ports: Smart ports, like those in Singapore and Rotterdam, are pioneering the adoption of quantum-inspired technologies to revolutionize container shipping operations.

PENNYLANE: Focuses on quantum machine learning and hybrid quantum-classical algorithms.

Photonic Systems: Use photons (light particles) to encode and process quantum information, especially valuable in quantum communication.

Post-Quantum Cryptography (PQC): Algorithms designed to secure systems against quantum computing capabilities.

Practice Programming: Experiment with quantum programming frameworks like Qiskit or Cirq.

Principles of Quantum Artificial Intelligence: Quantum problem solving and machine learning.

Programming Quantum Computers: Essential algorithms and code samples.

Programming

Pursue Formal Education: For those aiming for advanced roles, pursuing a degree or certification in quantum computing, computer science, or related fields is beneficial.

Quantum Acceleration: A rapid improvement in computational speed and efficiency achieved through quantum principles.

Quantum Advantage in Logistics: Using quantum computing to optimize supply chains, routing, and inventory management.

Quantum Advantage: Realizing unique computational capabilities that classical systems cannot achieve, particularly in areas like optimization and cryptography.

Quantum Algorithms for Genomics: Accelerating genome sequencing and analysis with quantum-enhanced algorithms.

Quantum Algorithms In Traffic Management: Optimizing traffic flows and reducing congestion using quantum solutions.

Quantum Algorithms: Specialized algorithms designed to exploit quantum mechanics, such as Grover's and Shor's algorithms.

Quantum Annealing: A method for solving optimization problems by finding the lowest-energy configuration of a quantum system.

Quantum Approximate Optimization Algorithm (QAOA): A quantum algorithm for solving combinatorial optimization problems.

Quantum Battery Research: Innovations in energy storage driven by quantum simulations for better performance and sustainability.

Quantum Bits (Qubits): Fundamental units of quantum information leveraging properties like superposition and entanglement.

Quantum Chemistry: The use of quantum simulations to study chemical interactions at a molecular level, enabling breakthroughs in material design and drug discovery.

Quantum Cloud Services: Platforms providing remote access to quantum computing resources via cloud infrastructure.

Quantum Coherence: The ability of qubits to maintain their quantum state over time, essential for reliable computations.

Quantum Collaboration Networks: Building interdisciplinary teams to address complex challenges in quantum computing.

Quantum Computation and Quantum Information: A foundational resource discussing progress and innovation in quantum computing.

Quantum Computing in Finance: Applications of quantum systems in portfolio optimization, risk analysis, and fraud detection.

Quantum Computing in Transport: Applying quantum solutions to optimize routing, scheduling, and fuel efficiency.

Quantum Cryptographic Protocols: Techniques like Quantum Key Distribution (QKD) that use quantum properties for secure communication.

Quantum Cryptography: Security systems utilizing quantum mechanics principles like the no-cloning theorem and entanglement.

Quantum Cyber Defense: Developing advanced security systems using quantum principles to protect against cyber threats.

Quantum Data Analytics: Enhancing data processing and predictive analytics with quantum tools.

QUANTUM DEVELOPMENT ENVIRONMENTS: Platforms like Xanadu and D-Wave Leap provide specialized environments for unique quantum tasks.

Quantum Disruption: The significant changes in industries and technologies driven by quantum advancements.

Quantum Drug Discovery: Leveraging quantum computing to accelerate the identification of effective molecules for medical treatments.

Quantum Ecosystem in Startups: The rise of innovative companies focusing on quantum solutions.

Quantum Ecosystem: The interconnected environment of quantum research, development, industries, and applications.

Quantum Energy Solutions: Applications of quantum simulations to improve renewable energy systems, such as solar panels and grids.

Quantum Error Mitigation: Techniques to minimize the impact of errors in quantum computations without full error correction.

Quantum Ethics and Responsibility: Addressing the societal and ethical implications of quantum advancements.

Quantum Financial Modeling: Revolutionizing risk analysis and portfolio optimization with quantum tools.

Quantum Future: The anticipated global impact of quantum computing on industries, research, and everyday life.

Quantum Gates: Fundamental operations on qubits, analogous to logical gates in classical computing, used in quantum circuits.

Quantum Grid Optimization: Revolutionizing renewable energy systems with quantum algorithms for enhanced efficiency.

Quantum Healthcare: Revolutionizing healthcare through genomic analysis, drug discovery, and personalized medicine.

Quantum Impact on Smart Cities: Enhancing urban planning, IoT networks, and energy management through quantum computing.

Quantum Impacts on Education: Transforming STEM education by integrating quantum computing into curricula.

Quantum in Telecommunications: Improving data transmission security and efficiency using quantum principles.

Quantum Industry Collaborations: Partnerships between academic institutions, governments, and corporations to advance quantum research.

Quantum Information Processing with Superconducting Circuits: A perspective on advancements using superconducting technology.

Quantum IoT (Internet of Things) The integration of quantum computing with IoT devices to improve security and efficiency.

Quantum Job Opportunities: Careers emerging in quantum software development, hardware engineering, and research.

Quantum Key Distribution (QKD): Enables secure encryption by leveraging quantum entanglement and preventing unauthorized interception.

Quantum Leadership: Pioneering efforts by tech companies like IBM, Google, and Microsoft in advancing quantum technology.

Quantum Leap in AI: The transformative impact of quantum computing on artificial intelligence and machine learning.

QUANTUM LEAP: How quantum computing will shape the future of technology, business, and everyday life.

Quantum Machine Learning (QML): Machine learning techniques enhanced by quantum computing for faster processing and superior results.

Quantum Mechanics: The theoretical minimum needed to understand the principles underlying quantum computing.

Quantum Monte Carlo Simulations: Advanced simulations using quantum processors to model uncertainties with high precision.

Quantum Networks: Communication networks leveraging quantum entanglement for secure and efficient data transmission.

Quantum Neural Networks (QNNs): Machine learning models enhanced by quantum computing for faster data processing and improved predictions.

Quantum Optimization in Healthcare: Leveraging quantum algorithms to improve personalized medicine, genomics, and drug discovery.

Quantum Parallelism: The ability of quantum computers to process multiple computations simultaneously using superposition.

Quantum Physics Simplified: Explaining foundational concepts like superposition, entanglement, and quantum interference.

Quantum Programming Languages: Specialized languages like Q# and Python libraries for creating quantum algorithms.

Quantum Readiness: Preparing organizations and individuals for the integration of quantum technologies.

Quantum Revolution in Transport: Optimizing logistics,

routing, and urban mobility using quantum computing.

Quantum Revolution: The ongoing technological shift driven by advancements in quantum computing, communication, and simulations.

Quantum Robotics: Exploring applications of quantum computing to enhance robotics' decision-making and learning capabilities.

Quantum Secure Communication: Using quantum mechanics to create unbreakable encryption systems for sensitive data.

Quantum Sensors in Medicine: Applications in diagnostics, imaging, and monitoring using quantum technologies.

Quantum Sensors in Navigation: Advanced applications improving accuracy and reliability in aviation, marine, and space travel.

Quantum Sensors: Devices using quantum properties to achieve highly sensitive measurements, such as in navigation or environmental monitoring.

Quantum Simulations in Energy Storage: Modeling materials for batteries and hydrogen storage to enhance energy efficiency and capacity.

Quantum Simulations: Techniques used to model complex systems in chemistry, materials science, and biology with unprecedented precision.

Quantum Simulators: Quantum systems used to model and study complex quantum phenomena in fields like chemistry and physics.

Quantum Society: The societal changes and challenges introduced by quantum advancements.

Quantum Supremacy: Achieved when quantum computers solve problems that are intractable for classical computers within a reasonable timeframe.

Quantum Talent Development: Preparing the workforce with skills and expertise in quantum computing.

QUANTUM THEORY MILESTONES: Planck, Einstein, Schrödinger. A series of groundbreaking contributions that redefined our understanding of the universe's fundamental laws.

Quantum Tools for Biochemistry: Revolutionizing protein folding and molecular interaction studies.

Quantum Transition: The gradual shift from classical to quantum technologies across sectors.

Quantum Urban Development: Enhancing urban planning and infrastructure with quantum-powered simulations.

Quantum Urban Planning: Using quantum simulations to optimize infrastructure, transportation, and resource management in cities.

Quantum Visionary Leaders: Pioneers advancing the field, such as Richard Feynman and David Deutsch.

Quantum Weather Predictions: Leveraging quantum simulations to improve climate models and weather forecasting accuracy.

Quantum Workforce: The growing field of professionals trained in quantum computing and its applications.

Quantum-Compatible IoT Devices: Integrating quantum technologies with smart devices for enhanced performance.

Quantum-Driven Economic Growth: The anticipated impact of quantum technology on global economies through innovation and efficiency.

Quantum-Enhanced AI: Artificial intelligence powered by quantum algorithms for faster training, optimization, and improved predictions.

Quantum-Enhanced Defense Systems: Strengthening national security through advanced quantum communication and cryptographic tools.

Quantum-Enhanced Weather Forecasting: Leveraging quantum algorithms to improve the accuracy of climate and weather models.

Quantum-Inspired Algorithms: Classical algorithms influenced by quantum principles to enhance computational performance.

Quantum-Resilient Encryption: Developing cryptographic systems resistant to quantum computing threats.

Quantum-Safe Cryptography: Cryptographic methods resistant to quantum attacks, such as lattice-based or hash-based algorithms.

Risk Management: Quantum-enhanced modeling tools revolutionize practices by offering unparalleled capital allocation and regulatory compliance accuracy.

ROI (Return on Investment): Businesses may hesitate

to invest in quantum computing due to its nascent state.

Shor's Algorithm: A quantum algorithm for efficiently factoring large integers, threatening classical encryption systems like RSA.

Start Small: Experiment with quantum computing by tackling proof-of-concept projects.

Steep Learning Curve: Quantum computing concepts can be abstract and counterintuitive.

Superconducting Qubits: Quantum hardware technology using superconducting circuits, often cooled to cryogenic temperatures for stability.

SUPERPOSITION: Beyond 0 and 1. One of the most fascinating and essential principles of quantum computing.

Surface Codes: Towards practical large-scale quantum computation.

THE QUANTUM LEAP: Why it matters. The combination of superposition and entanglement gives quantum computers their extraordinary potential.

Three Leading Technologies: Superconducting qubits, trapped ions, and photonic systems.

Traffic Flow Optimization: In a groundbreaking project, Volkswagen utilized quantum computing to optimize urban traffic flow.

Trapped Ions: A quantum computing technique using ions confined in electromagnetic fields to act as qubits, offering high precision and long coherence times.

Two Unique Features of Qubits: Superposition and entanglement differentiate qubits from classical bits.

UNCERTAIN ROI: Businesses may hesitate to invest in a technology still in its infancy.

VOLKSWAGEN: Optimizing traffic flow. Volkswagen has emerged as a pioneer in applying quantum computing to optimize urban traffic flow.

ZAPATA COMPUTING: Partnering across industries. Zapata Computing, a quantum software company, has partnered with various sectors to develop tailored quantum solutions.

REFERENCE

1. (Aaronson, 2013)Bernhardt, C. (2019). Quantum Computing for Everyone. MIT Press.
2. Nielsen, M. A., & Chuang, I. L. (2010). Quantum Computation and Quantum Information: 10th Anniversary Edition. Cambridge University Press.
3. Sutor, R. S. (2019). Dancing with Qubits: How Quantum Computing Works and How It Can Change the World. Packt Publishing.
4. Yanofsky, N. S., & Mannucci, M. A. (2008). Quantum Computing for Computer Scientists. Cambridge University Press.
5. Rieffel, E. G., & Polak, W. H. (2011). Quantum Computing: A Gentle Introduction. MIT Press.
6. Hidary, J. (2019). Quantum Computing: An Applied Approach. Springer.
7. Kaye, P., Laflamme, R., & Mosca, M. (2007). An Introduction to Quantum Computing. Oxford University Press.
8. Susskind, L., & Friedman, A. (2014). Quantum Mechanics: The Theoretical Minimum. Basic Books.
9. Gribbin, J. (2014). Computing with Quantum Cats: From

Colossus to Qubits. Prometheus Books.

10. Ferrie, C. (2018). Quantum Computing for Babies. Sourcebooks Explore.
11. Woody III, L. S. (2021). Essential Mathematics for Quantum Computing. O'Reilly Media.
12. Johnston, E. R., Harrigan, N., & Gimeno-Segovia, M. (2019). Programming Quantum Computers: Essential Algorithms and Code Samples. O'Reilly Media.
13. Wichert, A. (2020). Principles of Quantum Artificial Intelligence: Quantum Problem Solving and Machine Learning. World Scientific Publishing Company.
14. Zeng, B., Chen, X., Zhou, D. L., & Wen, X. G. (2019). Quantum Information Meets Quantum Matter. Springer.
15. Abott, D., Doering, C. R., Caves, C. M., Lidar, D. M., & Brandt, H. E. (2007). Dreams versus Reality: Plenary Debate Session on Quantum Computing. Quantum Information Processing, 6(6), 419–429.
16. DiVincenzo, D. P. (2000). The Physical Implementation of Quantum Computation. Fortschritte der Physik, 48(9–11), 771–783.
17. Krantz, P., Kjaergaard, M., Yan, F., Orlando, T. P., Gustavsson, S., & Oliver, W. D. (2019). A Quantum Engineer's Guide to Superconducting Qubits. Applied Physics Reviews, 6(2), 021318.
18. Simon, D. R. (1994). On the Power of Quantum

Computation. Proceedings of the 35th Annual Symposium on Foundations of Computer Science, 116–123.

19. Aaronson, S. (2011). The Limits of Quantum Computers. Scientific American, 298(3), 62–69.
20. Preskill, J. (2018). Quantum Computing in the NISQ Era and Beyond. Quantum, 2, 79.
21. Shor, P. W. (1997). Polynomial-Time Algorithms for Prime Factorization and Discrete Logarithms on a Quantum Computer. SIAM Journal on Computing, 26(5), 1484–1509.
22. Grover, L. K. (1996). A Fast Quantum Mechanical Algorithm for Database Search. Proceedings of the 28th Annual ACM Symposium on Theory of Computing, 212–219.
23. Montanaro, A. (2016). Quantum Algorithms: An Overview. npj Quantum Information, 2, 15023.
24. Childs, A. M., & van Dam, W. (2010). Quantum Algorithms for Algebraic Problems. Reviews of Modern Physics, 82(1), 1–52.
25. Lloyd, S. (1996). Universal Quantum Simulators. Science, 273(5278), 1073–1078.
26. Bennett, C. H., & DiVincenzo, D. P. (2000). Quantum Information and Computation. Nature, 404(6775), 247–

255.

27. Nielsen, M. A. (2003). Quantum Computation by Measurement and Quantum Memory. Physics Letters A, 308(2–3), 96–100.
28. Fowler, A. G., Mariantoni, M., Martinis, J. M., & Cleland, A. N. (2012). Surface Codes: Towards Practical Large-Scale Quantum Computation. Physical Review A, 86(3), 032324.
29. Ladd, T. D., Jelezko, F., Laflamme, R., Nakamura, Y., Monroe, C., & O'Brien, J. L. (2010). Quantum Computers. Nature, 464(7285), 45–53.
30. Gill, S. S., Cetinkaya, O., Marrone, S., Claudino, D., Haunschild, D., Schlote, L., Wu, H., Ottaviani, C., Liu, X., Machupalli, S. P., Kaur, K., Arora, P., Liu, J., Farouk, A., Song, H. H., Uhlig, S., & Ramamohanarao, K. (2024). Quantum Computing: Vision and Challenges
31. Wendin, G. (2023). Quantum Information Processing with Superconducting Circuits: A Perspective.
32. Mazumder, A., & Tayur, S. (2024). Five Starter Problems: Solving Quadratic Unconstrained Binary Optimization Models on Quantum Computers.
33. Upadhyay, S., Alam, M., & Ghosh, S. (2022). Architectures for Quantum Information Processing.
34. Gill, S. S., Marrone, S., & Ramamohanarao, K. (2024). Quantum Computing: Progress and Innovation. IEEE

Communications Surveys & Tutorials.

35. Wendin, G. (2023). Quantum Information Processing with Superconducting Circuits: A Perspective.

36. Mazumder, A., & Tayur, S. (2024). Five Starter Problems: Solving Quadratic Unconstrained Binary Optimization Models on Quantum Computers. arXiv preprint arXiv:2401.08989.

37. Upadhyay, S., Alam, M., & Ghosh, S. (2022). Architectures for Quantum Information Processing.

38. Gill, S. S., Marrone, S., & Ramamohanarao, K. (2024). Quantum Computing: Progress and Innovation. IEEE Communications Surveys & Tutorials.

39. Wendin, G. (2023). Quantum Information Processing with Superconducting Circuits: A Perspective.

AUTHOR PROFILE

OLADIPUPO JUDE MESOLE

Oladipupo Jude Mesole is a passionate technology futurist and thought leader, known for bridging advanced computing innovations with real-world solutions. His fascination with the next frontier of computation—**quantum computing**—inspired the writing of *Quantum Leap: How Quantum Computing Will Change the World*, a visionary book aimed at equipping readers to understand and harness the profound transformations this technology will bring to science, industry, and society.

Mesole has long been at the forefront of disruptive technologies. In 2013, he developed an AI-driven system to evaluate the risk of cardiovascular disease in humans—an early testament to his ability to apply emerging technologies to improve lives. With a diverse career spanning **banking, real estate, and consulting**, he has worked with major firms such as **KPMG, SAP, Hunters Group, and Prologis UK**, bringing strategic insight into digital transformation initiatives.

His academic journey reflects a commitment to both technological depth and ethical leadership. He holds a **Bachelor of Technology in Physics/Telecommunications** from **Federal University of Technology Minna, Nigeria (2006)**, an **MSc in Computer**

Engineering (Software Systems) from the **University of East London (2013)**, and a **Doctor of Theology** from **Christ Life Bible Institute and Seminary, New York (2020)**. He further pursued specialized training through postgraduate programs in **Data Analytics (Purdue University, 2021)** and **Applied Generative AI (Purdue University, 2024)**.

His certifications include

- SAP Business Process Consultant – SAP SD (Waldorf, Germany, 2013)
- Certified Data Analyst (USA, 2021)
- Certified Applied Generative AI Specialist (USA, 2025)

Currently a **PhD researcher in Organizational Leadership** at **Columbia International University,** USA, Mesole's research focuses on the **Physician's Perception of Ethical Leadership in the Deployment of Generative AI in critical sectors such as healthcare.**

As the global conversation shifts toward quantum innovation, Mesole's work bridges the technical and human aspects of transformation—advocating for **ethical foresight, inclusive innovation, and future-proof leadershi**p. With *Quantum Leap*, he invites readers into a world where unimaginable computational power could solve humanity's grandest challenges or deepen its ethical dilemmas, depending on how we lead.

www.ingramcontent.com/pod-product-compliance
Lightning Source LLC
LaVergne TN
LVHW050911210926
848482LV00012B/404
* 9 7 9 8 2 9 5 5 2 9 9 3 1 *